PODCASTING 101

A Complete Reference Guide to all Things Podcasting

TEDDY BURCH

innovativeink
PUBLISHING

Cover image © Shutterstock.com

www.innovativeinkpublishing.com
Send all inquiries to:
4050 Westmark Drive
Dubuque, IA 52004-1840

Print ISBN: 979-8-3851-0367-6
Ebook ISBN: 979-8-3851-00368-3

Published in the United States of America

CONTENTS

Acknowledgments vii

Chapter 1: Welcome! We're Glad You're Here 1

Let's Start with the Basics 2
Prepare Yourself: Successful Podcasting Isn't What You Think 4
Other Things to Prepare for 8
Where We're Going: A Breakdown of What You Can Expect Going Forward 11
Some Parting Thoughts 17
Chapter 1 Terms 18
Chapter 1 Checklist 19
Chapter 1 Questions to Ask Yourself 19

Chapter 2: Choosing Your Theme and Getting It Right.... 21

Moving Forward and Laying the Groundwork 24
The Psychology of Logo Design 30
Choosing the Correct Font 34
Choosing the Right Shape for Your Logo 36
Where to Start in the World of Software for Creating Logos 38

Chapter 2 Terms 41
Chapter 2 Checklist 42
Chapter 2 Questions to Ask Yourself........ 42

Chapter 3: Setting Up Your Studio 43

What You'll Need 43
Getting the Proper Cables 47
Headphones: Eeny Meeny Miny Moe........ 51
Audio Boards: Now We're Getting to the Fun Stuff........ 53
Soundproofing: Is This Part Necessary? 55
So, Now What?........ 57
Full-Production Studios: It's Time to Get Serious 57
A Brief Rewind 66
Chapter 3 Terms........ 67
Chapter 3 Checklist 68
Chapter 3 Questions to Ask Yourself........ 68

Chapter 4: Show Prep, Episode Mapping, and Script Writing 69

Script Writing 74
Scripting the Intro........ 76
Chapter 4 Terms........ 79
Chapter 4 Checklist 79
Chapter 4 Questions to Ask Yourself........ 79

Chapter 5: Interviewing and Recording Your Podcast 81

Recording Your Podcast 83
The Intro........ 84
The Interview—One Host and One Guest 84
The Interview—One Host and Multiple Guests 86
The Solo Show........ 89

Some Tips on Becoming a Better Interviewer 89
A Few Tips on Having a Better On-Air Appearance 91
An Audio Recording Checklist..... 91
An Audio/Video Recording Checklist..... 92
Chapter 5 Terms 94
Chapter 5 Checklist 94
Chapter 5 Questions to Ask Yourself 94

Chapter 6: Audio and Video Editing (With Geoffrey Burch) 95

Audio Editing—Let's Start Here 95
Postproduction 97
Video Editing 104
Trimming 109
Moving Clips in Your Timeline Panel..... 110
Adding Graphics 110
Video Effects 110
Exporting 112
Chapter 6 Terms 114
Chapter 6 Checklist 115
Chapter 6 Questions to Ask Yourself..... 115

Chapter 7: Social Media Promotion and Distribution ... 117

Five Myths That Simply Aren't True About Social Media Marketing 118
Five Tips for Better Social Media Promotion 120
Where to Store Your Podcast 123
Let's Bring It All Around 123
Chapter 7 Terms 126
Chapter 7 Checklist 126
Chapter 7 Questions to Ask Yourself..... 126

Chapter 8: Copyright Law and the FCC 127

Three Myths 128
Copyright Law Q&A 128
Federal Communication Commission 134
Chapter 8 Terms 137
Chapter 8 Checklist 137
Chapter 8 Questions to Ask Yourself 137

Chapter 9: Keeping It Going 139

Staying Motivated 142

Chapter 10: Troubleshooting Your Problems 145

Audio Issues 146
Video Recording Issues 148

Glossary 151

ACKNOWLEDGMENTS

I would like to take a brief opportunity to thank a number of folks who helped me in this long and formidable journey of writing a textbook. I have the deepest gratitude for my wife Dena who for months dealt with me being locked away in a quiet room to spend endless hours putting this book together. Thank you! To my brother Geoffrey, who is busier than probably all of us, and yet he made time to write an extremely important portion of Chapter 6, again a major thank-you! To Innovative Ink for seeing the importance of there being a book just like this one for all of us podcasters! To Bill Wilson for his knowledge of FCC things. To Noah Mears, who as a college student is making the very most of his time and is reaching for greatness. There is no doubt he'll accomplish this . And finally, to all the current and future podcast students—together we're going to rock the podcast world, and this book will be our map to do just that!

1

Welcome! We're Glad You're Here

Hello! We're all glad to see you. Since you're here, there's something we have in common: a love for podcasting. Maybe you're here because you have this wonderful idea for a podcast and it's time to take that idea to the next level and begin putting together your show. Perhaps you're here because you've already got a podcast, but there's something missing, something that needs to slightly improve to make your production ready for the next level. Or maybe, you are in a university course on podcasting and the excitement is building and you're ready to jump in with both feet and get started. No matter the reason behind your reading this book, a way to improve your podcast is here.

LitDenis/Shutterstock.com

IMAGE 1-1 Welcome, we're glad you're here

As you may have read in the table of contents, this reference manual contains all the information you would need in order to set up the simplest of audio-recording studios and to build elaborate, full-production studios capable of audio/video recording and live streaming. Think of this book as more of a "How To" rather than a typical college textbook. Sure, there are questions associated with each chapter, there are exercises you are encouraged to take part in at the end of each chapter, and who knows there may be a test over this material awaiting you down the road. Either way, this is *the* reference manual for your project. Let me encourage you to come along on the journey of taking your ideas and your passions beyond, "I should do a podcast on ...," on to creating that podcast and sharing it with the world. Who knows, the world just might want to hear what you have to say.

Let's Start with the Basics

What exactly is a podcast? Well, I'm glad you asked. In the simplest terms, a podcast is a digital file, usually an audio file, that has been uploaded onto the Internet so that a listener can download the file onto a digital device for listening. In so many words, that is a podcast. Episodes can be released individually, or creators may wish to have a series released at once so that a listener can download and listen as they choose. A podcast can consist of a host, a host with a guest, a host with a cohost, or a host and cohost with guests.

When a podcast has a host, or a host and cohost, and one or more guests this is known as an interview-style podcast. It can be very informative, but there is a catch—the guest. We'll get into that a few pages from now. A conversational-style podcast, on the contrary, consists of a host and cohost and usually never has guests. This is an easier format for the hosts in that there are no secondary obstacles; however, rhythm among hosts is critical. Another type of podcast is a solo podcast. This is just as it sounds, and the host is solo without a cohost and without guests. Often I have students choose this format because prep time is minimal, and there are virtually zero obstacles to overcome. I've had students start the semester down this path and about halfway through the semester ask to change formats. Why? When you attempt a solo podcast, you are the show. The entire show. You have to keep the conversation going solo, and for some of us this may sound easy, but it's actually the opposite. It is hard to have episode after episode of quality solo content.

Dean Drobot/Shutterstock.com

IMAGE 1-2 You'll have to decide which format is best suited for your podcast

You may find yourself in a unique situation where you are highly intelligent in, say, stock investing, or mass multiplayer online gaming, or even gardening. The list is endless, but if you fall into a category of knowing a lot about one particular domain you may want to consider an educational-style podcast. An educational podcast is similar in style as the abovementioned ones; however, the host and/or cohost would focus on one particular element of their domain and each episode focuses on a different, related topic.

Podcast theater is a growing format in the world of podcasts. Are you a fiction writer? Or maybe a fiction writer in the making? Then this is an area to consider. And it's just as it sounds: it's theater. This is where the host, or both the host and cohost, would have an opportunity to share their short fiction stories. I've had theater majors who have taken my class simultaneously for the sole purpose of bringing their fiction to life. This is an area where currently the podcast world isn't saturated and could be just the right fit for the right writer. One thing to be watchful for in this category is to avoid long, novel-like episodes. That is reserved for audio books, which is an entirely different area of communication.

Another popular podcast format is known as the panel podcast format, and like the name implies the host would have a panel of guests who are well-informed in a particular domain and the panel discusses the topic chosen by the host. Very popular for politics, movies, and religion. You know the topics where everyone has an opinion and theirs is always right. Due to the nature of this format, topics for discussion can vary from a wide range of issues and allow for the host to be versatile in their topics, and remain very timely and current.

Finally the sixth and final podcast format we'll discuss in this chapter is similar to podcast theater and is, you guessed it, nonfiction podcast. This is a podcast type that uses nonfiction topics and discusses and informs listeners on a wide range of topics. Unlike the educational format, this format allows the host to switch topics for each episode. Perhaps, in one episode, the host may want to inform their listeners the background of where their eggs in their refrigerator come from, while the next episode may inform listeners how the energy to run that refrigerator was generated. You could literally pick a Wikipedia topic at random, become highly informed on the topic, and create your podcast about that topic. This, as you may imagine, is a highly popular format, one where the real character of the host has the opportunity to come out.

Prepare Yourself: Successful Podcasting Isn't What You Think

Ask yourself, *How do I podcast?* Yes, we are going to use the term podcast as a noun, as a verb, and who knows, maybe even as a preposition. But to get back to that question, what's your answer? Is it while you're driving? Is it while working through those seemingly endless hours of college homework? Maybe it's while folding towels while the dinner is cooking in the oven. Whatever your answer, how that program ended up on the streaming channel of your choosing is not how you're thinking. More on this in just a few.

Now, let us think about our favorite podcasts. At the time of writing, comedy, true crime, news, and society and culture were some of the most popular genres. You might be listening to something similar now. Think about some of the most popular shows in the industry, *The Daily*, *Stuff You Should Know*, *The Joe Rogan Experience*, all are good podcasts that meet the needs of their listeners, but once again, it isn't what you think.

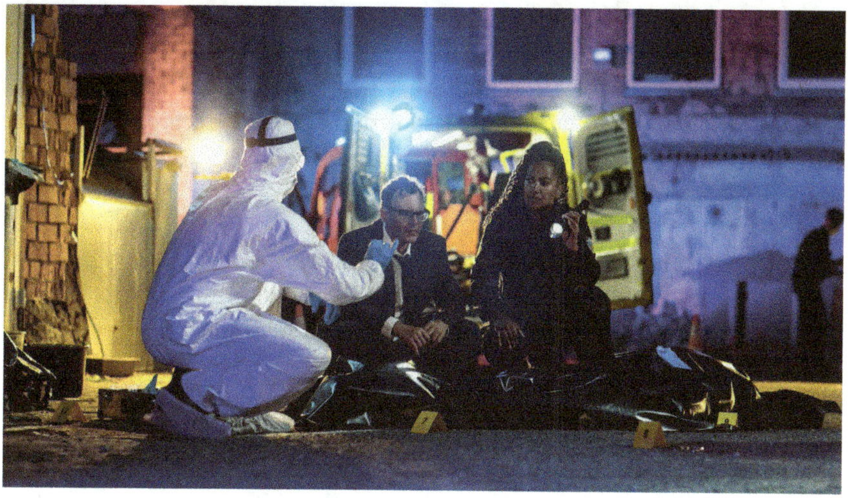

Gorodenkoff/Shutterstock.com

IMAGE 1-3 True crime is just one of the many popular podcast genres

You see, most listeners tune in to a podcast because of the way the program makes them feel. They want to be reassured that their political opinions are valid and have merit, or they want to laugh out loud and forget about the history paper that is due in less than 24 hours. Or they simply wish to be informed or may be just a little spooked. These are all valid reasons for tuning in, and the host of the podcast had better give their audience what they've come for, and the above listed shows and their hosts do an outstanding job of delivering the goods. However, what we need to do is to begin shifting gears, to begin flipping that switch up there in our minds. The switch somewhere in between "What's for dinner?" and "Did I turn the coffee pot off?" The switch that begins to tell ourselves that we are no longer consumers. We are not here to be entertained or to have reasons to dislike the host. No, we are here to entertain, we are here to enlighten, we are here to inform. *We* are the podcasters.

And just as some of us are shaking our heads up and down slightly and saying yes, or the feeling is growing to the point of being ready to run through the paper banner to take the field for the big game, let me bring around little reality to it. What we need to prepare ourselves for are a couple of big F's: failure and frustration. You see, that's the part you never hear when you're cruising through afternoon traffic, hitting all the green lights with your favorite podcast playing. Whoever you're listening to has received those F's. Probably dozens and dozens of times over. The difference between that podcast you're going to listen to next and the hundreds of thousands of great ideas that never got off the ground is that the host you're listening to didn't quit; they didn't allow the F's to extinguish the drive and zeal every newbie podcaster has. They didn't bow to rejection or quit when the episode didn't actually record. Nope, those podcasters have what I call the main ingredient for any successful podcasting journey: perseverance.

Got a wonderful idea for a podcast? Great. Have the perfect voice for podcasting? Wonderful. Have lots of contacts that will be outstanding guests for your show? Swell! While all of those are important (you're going to need each of those) they are all secondary ingredients. If you don't have perseverance or, as *The Oxford English Dictionary* would state, "persistence in doing something despite difficulty or delay in achieving success," let me take this opportunity to beg with you— no, plead with you— to take your wonderful ideas and your creative thoughts and do something else with them. If you do not start out on your podcast journey with big ole heaping, helping dose of perseverance, you will accomplish one of the two F's, and you will fail.

Rather, what you need to do at this moment, this very moment, is to tell yourself, teach yourself, convince yourself that you will persevere and overcome the many obstacles you are going to face on the road to a successful podcast. Become comfortable with the idea that you're not going to quit, and you're not going to retreat when that *perfect* episode of yours (at least in your mind) have got 14 listens and 8 views. Prepare for it now. That's the frustrating part of the two F's. Prepare for it now because it's going to happen. It was Eddie Cantor who said, "It takes twenty years to become an overnight success." Let's hope it doesn't take you 20 years; however, you are not going to have one million-plus listens by this time tomorrow. Prepare for it now.

Are you still with me? Haven't slammed the book shut and told the author what they can do with their opinions? Good, because I promise this book is still *the* reference manual for successful podcasting; there are, however, a few more loose ends that need addressing before we turn up the microphones and turn on the recorders. And here is the next one ... are you ready for it? OK, here it goes— your concept for your podcast is incomplete!

That's right; your concept or the topic of your podcast is incomplete at best. No matter if the idea is 5 minutes old, or you've been rolling this idea around for weeks now, it is incomplete at best.

Gustavo Frazao/Shutterstock.com

IMAGE 1-4 The most important tool you can have on your podcast journey is perseverance

Maybe this is where you place a bookmark, or make note of where you are in the book and walk away for a while. No problem, we get it. Or if you'd like to persevere (see what we did there?); let's take a look at that statement again. Your concept for your podcast is incomplete. Notice I didn't say your concept is terrible, I didn't say your concept is saturated throughout the podcast community. I didn't even say that your concept lacks depth and interest. What I did say is that your concept is incomplete. That is to say, your concept may know where it's beginning, but it has no idea where it's going. Don't believe that? Go back and listen. Go back to any of your favorite podcasters, back to their beginnings, and listen to, truly listen, for nuances of language, the questions asked, the range of guests. Now listen to one of their most recent episodes. What difference do you notice? They too once had an incomplete concept.

So how do you prevent this? How can one avoid such incompleteness heading into the beginning of their podcast? Honestly, you can't. I once was driving in the back roads of middle Tennessee, a beautiful place by the way, listening to a podcast about podcasting, strange enough. Anyhow, the host and cohost were about to record their 400th podcast later that day and they were asked how long did it take to develop their character and their voice for their podcast. Their reply? Thirty to 40 episodes before they were comfortable with themselves, their cohost and their character. I was not expecting 40 episodes. We are going to discuss more about concepts and themes in Chapter 2; however, between now and then I'd like you to begin developing your podcast theme. Allow it time to evolve, to mature somewhat. Part of the incompleteness is going to be obvious from your beginning stages until your first recording. Be comfortable with that; let your concept come to you, and force nothing.

Other Things to Prepare for

I've taught this course for several years, and there are some consistent issues that pop up each semester. Some of these issues you can control, which is great because you can decide now or later that the issues you can control should be the least stressful of the problems you are inevitably going to encounter. Then you'll have issues that you simply cannot control and these are the ones that are going to set you back in the production of your podcast. These can also be the ones that bring on frustration the quickest.

Arsenii Palivoda/Shutterstock.com

IMAGE 1-5 Sometimes guests can be the most difficult to schedule and get into the studio

So let's start here, let's start with the biggest problem that is out of your control. Ready? Your guests. That's right; the friends, family members, coworkers, and longtime colleagues, they're going to let you down sooner than later—prepare for it now. What's that you say? So and so going to be a no-show? Impossible. Actually it's not, and it's more common than I'd like to admit. Sure, their intentions are good, and some of them truly want to help you get your show off the ground. But sooner or later— probably sooner— so-and-so is going to get sick the day of recording, or they'll forget all about it and have made other plans, or as I had a recent student experience, they will ghost you while you are texting directions to get to the studio. Remember, your podcast is not as important to them as it is to you. Make preparations for backup guests, as at some point you're going to need them.

Here's another thing in relation to your guests; make sure you can get them. By that I mean I've had some students with outstanding concepts for their podcast. *Unsolved Mysteries throughout Tennessee History*, *Ghost Hunting*, and *The Truth about Extraterrestrial Life*. All great topics; all answer the questions on deciding on a topic (more on this in the next chapter), and all seem interesting. What's the problem? Well, until the day comes and your podcast reaches the point where guests are calling you in an attempt to make an appearance on your show, you need to make sure your guests, an authority in their respective fields, are available. For example, I'm working on a project at the moment, it is a podcast titled *Let's Go Smoke*. I know what you're thinking but it's not like that. This podcast is all things BBQ. Why? Because I love BBQ, I have enjoyed a number of smokers over the years. I have studied and practiced just about every local wood paired with just about every cut of meat, and have found some to be close to perfection. Another reason, I am surrounded by fellow BBQ enthusiasts. There must be a dozen restaurant owners in the area, and several pitmaster champions live in the same city as I do. I would have 20+ guests on the list before the first script has been written. It might be a good idea for you to do the same.

There is something else that is out of your control and at some point it's bound to happen: equipment failure. In Chapter 3, "Setting Up Your Studio," we will discuss all things equipment; however, before then I'll just suggest you learn all you can about your recording equipment. At this point in the game, you are a one-person crew. You are the talent, the producer, the editor, the audio engineer, the promoter, and the content writer. As we break each of these down, all are vital to a well-produced product; however, fully knowing the function of your studio equipment is arguably the most important. The question isn't whether something will not work in the studio but rather when. And usually it's when you have your guest sitting on set, under the lights.

IMAGE 1-6 You're the boss of your podcast, you are also most likely every other member of the production crew as well

How about some things you can control? Here's one, your pace. I don't mean the pace of your show, although we will address that in Chapter 4. No, right here I mean the pace of the approach to your show. As of now, I hope you are becoming familiar with the idea there's a lot to do before you ever record. Some of you are right now thinking I can record two or three shows a week, and release every other day. I would say, don't do that. Pace yourself. You are going to have shows where you'll believe the hand of God is guiding your every move, recording goes great, editing is a breeze, and listeners are at an all-time high! Then you'll have shows where the guest would be a dud, your chemistry is terrible, and editing turns into an all-night event. Most every show will fall somewhere in between. Pace yourself. Don't let your concept flame out before it gets a chance to find itself.

A couple of more things to prepare for right now. First, editing is going to take longer than you'd expect. Probably a lot longer than you'd expect. As we'll discuss in Chapter 6, editing is probably going to be the single longest activity in the entire process. One day, when you're able to expand your team and have a dedicated audio/video editor, you'll have more time for other tasks, but until then you'd be the one manning the editing bay. Depending on which software you use—and there are a lot of them out there—there is a learning curve to each one, and mastering the tools and techniques of these software programs takes time. Again, each step in the process of producing your podcast is important, but a rushed editing job sounds and looks bad. Doesn't sound good, doesn't look good—well, that equals a poor product and almost a guarantee of nobody listening.

Second, social media promotion is an ever-changing set of tools that you must learn to comprehend and prepare for daily use. Chapter 7 will provide some great tips and tricks to get the most mileage out of your social media promotion time. Just plan on using the medium often; don't be afraid to migrate to new social media platforms, and don't underestimate the power and the reach of every platform. In fact, almost everyone reading these words right now have been active on multiple social media apps on the very day.

Where We're Going: A Breakdown of What You Can Expect Going Forward

In the upcoming chapters, you are going to become informed on all things podcast, and in just a few short moments you are going to get a quick breakdown of each chapter. But before that, I'd like to share with you a couple of components that each chapter is going to emphasize so that you get the most mileage out of each chapter. At the end of each chapter, with the exceptions of Chapters 9 and 10, you are going to have a list of terms that were introduced in that chapter's reading. I know, this isn't exactly groundbreaking stuff, but the point of these terms is not to be able to pass an exam but to know the ins and outs of complete podcasting. For many of us, it'll be like learning a new language. Having a complete and full understanding of podcasting and recording language is vital.

Each chapter is going to have a checklist. These checklists aren't intended to be checked off as you read the contents but rather, if you would think of this book as your reference manual, are intended for you to refer back to specific concept, mark them off as the task is accomplished, and prepare yourself for a successful launch of your podcast. You can skip this portion of the book if you like; however, in so doing you are introducing yourself to a whole new level of unforeseen failure and frustration, which would urge me to ask, Why would you do that? I digress; the purpose of the checklist is to assure you are not overlooking any detail, no matter how small, and you have prepared yourself for success.

lassedesignen/Shutterstock.com

IMAGE 1-7 Like most other endeavors in life, knowing the language goes a long way toward success

Finally, each chapter is going to have a set of questions. We all dread the questions at the end of the chapter. It feels like meaningless page flipping to hunt for bold words, or recall that sentence you read that may hold the correct answer. Yeah, that's how I feel as well. So these questions are different. These are questions that you will be *asking yourself*. And if you are being honest with yourself—and I strongly encourage that you be—you cannot get these questions wrong. The purpose of these questions are for you to set little mile markers along the journey with the intentions of you honestly answering where you are in the journey to setting up your podcast. I know a lot of us are ready to begin recording right now. You have this outstanding message that the world needs to hear. And it's important to get recording today, even right now! Take a deep breath, The simple truth is, we're not ready for that just yet. Let's get our foundation set first, and then, I promise, your message will be heard.

Pheelings media/Shutterstock.com

IMAGE 1-8 Take a deep breath, your podcast journey is about to begin!

So let's take a quick peek into what the upcoming chapter has in store for us. But first, I've got to include a very important statement that is (1) required and (2) 100% transparent. Here it goes:

In the upcoming chapters I do not promote the use or the purchase of any proprietary software programs, computer manufacturers, studio recording brands or manufacturers, or any streaming services. I am simply listing what I use to teach and to record. At no point has there been or will there be any paid-for promotion for any brand of manufacturer within this book. These are brands that I have chosen to use.

OK, now that that's done, let's get started with what you can expect from each chapter.

Chapter 2—Choosing Your Theme and Getting It Right

This is a vital part of the process. Earlier I encouraged you to begin thinking about your concept for your podcast. How are you doing with that? In Chapter 2, we'll cover all the steps in the process of creating the look and feel for your podcast. This cannot be overemphasized enough, and this step is a crucial first step in your success. Prepare to take as much time as possible to get this one right. The truth is you are going to need to have this one completely thought through by the chapter's end. You've got to find your niche, and researching other podcasts to see where you could offer a means of filling a void is time well spent. In fact, I don't believe you can spend too much time on this step. It's vital to the beginning of your podcast.

Chapter 3—Setting Up Your Studio

I've had the privilege of designing, building, and finalizing multiple studios. You'd be surprised at how spaces that started as broom-and-mop closets, or even as old storage rooms for radio stations that had FCC regulation books dating back to the early 1970s stacked to the ceiling—with a little imagination, and hard work—can be converted into your podcast studio. While Chapter 3 will address spaces, the major focus is on identifying the equipment you'll need for a successful setup. You'll learn what TRRS cables are and what their functions are, you'll get a full rundown on unidirectional, omni, and bidirectional microphones, you'll even get familiar with switchers, audio boards, and how to record your podcast for a seamless transition into editing. While Chapter 3 covers a lot, we'll approach it one bite at a time, making it digestible for anyone wanting to set up their studio.

TZIDO SUN/Shutterstock.com

IMAGE 1-8 We're going to get into the depths of studio design

Chapter 4—Show Prep, Episode Mapping, and Script Writing

Script writing for podcast is different from broadcast or radio. We'll cover the dos and don'ts, we'll practice our intros, and we'll discuss the power of planning the blueprint of each episode. There are some variations depending on your format and we'll take a look at each. It is right here where I believe it is important to state that you cannot overdo your research and your preparations for each episode. In fact, one of the worst practices you can get into is underpreparing for each episode. Even if you choose an educational format podcast, and you are the smartest person in the room in regard to your topic, do your research, be overprepared. I promise, your listeners will pick up on it.

Chapter 5—Interviewing and Recording

By the time we reach this portion of the journey, we'll have finally arrived at the point where you can hit the record button and begin your masterpiece. In Chapter 5, we'll discuss interviewing techniques—when to ask questions and when to listen and how to identify key moments in the interview for moving it forward for your listeners. You'll become familiar with the difference between a good interview and a bad one. And you'll be able to prepare yourself for avoiding the latter.

Chapter 6—Editing

This is the portion of the process where the real gold happens. I'm a big believer in not creating work for yourself. If you can address and fix a problem before it's recorded, do it. While we'll address sibilance and plosives and how to avoid them in Chapter 6, it is inevitable that they're going to happen. In fact, a door closing, a guest coughing, or simply rocking back and forth is going to have an impact on consistent quality. Chapter 6 addresses and deals with all issues to produce the best quality podcast possible. Remember, you are a one-person production crew at this point, and being able to master the art of editing is crucial. Chapter 6 has a lot of depth to it, but again you need not worry; we'll attack this chapter just as we have the previous ones, one bite at a time.

Chapter 7—Social Media Promotion and Distribution

Well, this is a big one. What ought one do with that finished podcast? In Chapter 7, we'll go through a step-by-step process of how to maximize your distribution. We'll cover ways to begin getting your podcast where people can find it. We'll cover analytics, and monetization of your programs. There will also be some of the greatest in the profession of social media management join us for supplying us with some tips, tricks, and tried and true methods for getting your podcast in front of others via a multitude of social media channels.

One parting thought before we move on to the next chapter summary, of all the material covered up to this point, this is the section that needs attention every day. Let me say that one more time, this is the section that needs attention every day. Want a successful, popular podcast? It takes attention every day and your social media strategy needs to reflect that.

Chapter 8—Copyright Law and the FCC

We gotta stay safe and we gotta stay within the law. Chandler and Munday (2011) define "media law" in *A Dictionary of Media and Communication* as,

> The legislation through which governments regulate the mass media. It includes issues of censorship, copyright, defamation, broadcast law, and antitrust law. In democracies, media law is seen as a balancing act between two conflicting principles: freedom of expression and constraints laid down in statutes of common law, as in issues of defamation and the national interest.

Media law may or may not be your strength, but you've to stay on this side of the law, you need to steer clear of copyright infringement, avoid defamation, and essentially keep yourself out of trouble. We'll get neck deep in media law, and while that may have you thinking twice about launching your podcast, don't let it. Sure, it can be intimidating, but just like I've mentioned earlier, we'll get through it one step at a time. How do you eat an entire supreme pizza with extra mozzarella cheese? One bite at a time, of course, and that's just how we are going to approach media law.

Chapter 9—Keeping It Going

So many times I've witnessed good talent begin an interesting podcast, and they're having fun, and getting views, and all's right with the world. Then after a certain number of episodes, dread sets in. They dread the process, they dread the editing, and they dread the social media postings. What usually follows is the show withering like a desert flower in the scorching sun and the world never getting to experience what could have been a great show. Chapter 9 will provide some ways to avoid this. We'll do some exercises in preventing burnout and introduce some ways to use your talents elsewhere to further develop your abilities while working with others to help prevent podcast fatigue.

Chapter 10—Troubleshooting Your Problems

The last stop on our reading journey is Chapter 10. Here, we'll cover some of the most common problems you'll run into along the way with your recording studio, and hopefully explore some solutions.

Jasni/Shutterstock.com

IMAGE 1-9 Proper troubleshooting will save you time and frustration

Some Parting Thoughts

To this point, we've covered a lot of ground in a few pages. I hope you are preparing yourself in several different ways for success. There is plenty to do in the process of podcasting and now is the time to do it. Remember, a few pages back, when I mentioned that the end of each chapter would have some terms to know and a checklist? Well, here we are at that point of for Chapter 1. Remember, when thinking about the following checklist, it's not something you do at the conclusion of the chapter but rather it's a list you check off as you accomplish the specific task. We'll get to that checklist in a few, until then, here's your terms.

Chapter 1 Terms

Cohost The second of the two individuals leading the podcast.

Conversational-style podcast Consists of a host and cohost and usually never or very rarely has guests.

Educational-style podcast The host and/or cohost focus on one particular element of their educational topic and each episode focuses on a different, related topic.

Guest An knowledgeable individual brought onto the show to speak about a specific topic.

Host The individual leading the podcast.

Interview-style podcast When a podcast has a host, or a host and cohost, and guests.

Nonfiction podcast This is a podcast that uses nonfiction topics and discusses and informs listeners on a wide range of topics. Unlike the educational format, this format allows the host to switch topics for each episode.

Panel podcast format The show's host has a panel of guests who are well-informed about a particular topic and the panel discusses the topic chosen by the host.

Persistence Doing something despite difficulty or delay in achieving success.

Podcast A digital file, usually an audio file, that has been uploaded onto the Internet so that a listener can download the file onto a digital device for listening.

Talent The person, or persons, in front of the camera hosting the podcast.

Theater podcast This is where the host or the host and cohost have an opportunity to share their short fiction stories.

Two F's Failure and frustration (prepare for them; they're just around the corner).

Vodcast A digital file similar to a podcast; however, there is a video component accompanying the audio file, to be watched as well as listened to.

Chapter 1 Checklist

- I've read Chapter 1 and have a firm understanding of its context.
- I have an idea to this point of the format I'm thinking of for my podcast.
- I know why I have chosen that format.

Chapter 1 Questions to Ask Yourself

(Be honest with yourself, you might as well be.)

1. How do I podcast?
2. Why didn't I choose the other types of podcast formats?
3. At this point, what percentage am I truly dedicated to seeing my podcast through?
4. Am I OK with experiencing the two big F's?
5. Am I fully prepared to put in the time to edit my podcast to make it the best it can be?
6. Am I fully prepared to put in the time necessary for social media promotion to give my podcast the best chance possible to be successful?
7. Do I understand there is no easy path to a successful podcast?

2

Choosing Your Theme and Getting It Right

Well. hello there, you made it back! Since you're here it looks like you are going on the second leg of this journey and are serious about this podcasting thing. Great. OK, well, this is the portion of the journey where things get serious. This is the part where things begin to come to life, this is the part where you are going to choose you theme, your format, and the feel of your podcast.

There's a lot to this process, that is, if you want to get it right for the long-term. So, where do we start? Before we design logos, or think about color schemes, we need to decide on the format of our show. Think back to Chapter 1. We discussed six different formats, and typically, all podcasts are going to follow one of those formats. Table 2-1 lists the format, the pros and the cons of each style.

TABLE 2-1 Pros and cons for each style of podcast

FORMAT	PROS	CONS
Interview-style podcast	- A very versatile format for a podcast - Allows the host to have a wide variety of topics - Because it is conversational style editing times can be reduced.	- You will have to be versatile in your scheduling; remember the problems that can and will arise with guests? - This is arguably the single most popular format in podcasting. Want to stand out from the crowd? You're going to need a unique angle, with a unique delivery.

FORMAT	PROS	CONS
Conversational-style podcast	- Should be in a very natural environment, so preshow planning is reduced - With a natural chemistry between host and cohost, topic is always flowing - Allows for listeners to naturally feel part of the conversion - With two hosts generating content ideas, it should naturally allow for a wider range of topics.	- Because so much of the show depends on chemistry and natural conversational flow, it has to exist between host and cohost. - It is vital that both hosts understand they are equal in rank and expected to have equal speaking time.
Educational-style podcast	- Depending on the topic you choose, it is likely your audience is going to become invested in your teachings, and likely will continue to come back for more episodes. - The more episodes you record, the more you'll learn about your topic. - A secondary means of monetization exists in this format because you can create and sell videos, mini courses, or even write a textbook.	- Must choose your topic carefully if you are in for the long run. Even a topic you know a lot about has a limited number of episodes. Be sure and choose a topic you can get as many subtopics as possible. - There has to be an audience for your topic. Just because you know a great deal about a particular topic doesn't always equal an audience that cares.
Theater-style podcast	- Excellent opportunity for self-expression - Excellent opportunity to further develop as a fictional writer and even as a performer. - Probably the least populated format in podcasting at the time of this publication.	- Easily the most difficult format as far as planning and show prep. - The podcast is only one part. The host of this style must also be a writer and actor to pull this off successfully.
Panel-style podcast	- The best format for getting into the middle of current topics - Gives the host the opportunity to approach controversial topics from both sides - Given its versatile nature, this format has a long-term ability in that topics are virtually endless.	- Doubles or triples the opportunity to be stood up by guests - Can create more editing time - Requires the host to throw a wide net in order to get experts in the field of chosen topics

FORMAT	PROS	CONS
Nonfiction-style podcast	- Isn't limited to any one topic - Could have the best long-term ability of any of the podcast formats discussed in this book - Allows the host to explore topics resulting in learning many new things	- Must be willing to research many topics for show preparation - Much like the panel style, the host must be willing to reach out to large field in order to get knowledgeable guests for each topic.

So, this is a critical step in the process. Which format are you choosing? Hold on for just a second. Before you say it out loud, I want to think about answering one question: Why? Why did you choose that format? Better yet, why didn't you choose another format? While this is somewhat testing your decision, I want you, no wait, I need you to be sure. Are you sure that you have chosen the right format for you and that you are setting yourself up to succeed? Take the opportunity and the time to run through these questions in your mind. When you have answered these questions, continue on with this chapter. I promise, the rest of the content in this chapter will make more sense once you've arrived at your format, and you can truly answer these questions.

ImYanis/Shutterstock.com

IMAGE 2-1 Pondering which format your podcast is going to be is time well spent

Moving Forward and Laying the Groundwork

OK, so you've arrived at answers to all those important questions. That is great news. So what's next on the agenda? I'm glad you asked. What we need to do now is to begin developing your concept. Take your time with this one. Now we know your idea, no matter how well-thought-out, is going to be incomplete at this point. Refer to Chapter 1 if you need a refresher. Anyhow, we need to begin putting together the parts of a well-thought-out concept. Start by asking yourself a few more questions. *First, why do I want to get into podcasting?* Be honest. *For the genuine fun of podcasting?* If yes, that's awesome because podcasting is fun, and it's fun to share. *Is it because I have a message that needs to be shared?* That too is cool, and a perfectly good reason for getting into podcasting. Maybe you're getting into podcasting because you want to make money while enjoying what you do for a living. To be honest, that's why I got into teaching higher education. I needed a job, I love teaching students, and having summers off is a nice bonus. If making a paying career out of podcasting is your objective, that is outstanding! It'll probably give you more motivation to power through when things get monotonous.

The next question you should ask yourself is what. *What do I want to use as my topic?* It should be something you have a great passion for, something that is going to excite you to talk about. A brief example of how not to choose a topic. In graduate school, I had to take analysis of variance (ANOVA); if you've been there, you'd know what I'm talking about. I have nightmares about statistical formulas that look like roadmaps written in a foreign language. Not sure I've ever had a more difficult 3-month period in life. I would never, ever entertain the idea of a statistics-themed podcast. You, on the other hand, may get overly excited, with your heart pounding to the thought of regression and *t*-tests. If so, you should choose this as your subject, and you'd have very little competition. Joking aside, you need to narrow down your concepts to just a few, and then don't move too fast. Allow the proper topic to develop in your mind and then you'll know you're on the right path.

Starting to form your concept? Excellent! Now it's time to begin our research. By this I mean who else is doing this? Are there 48,755 other podcasts exactly like the one you are creating? If so, you need to keep researching. That is, your concept isn't terrible; in fact, the opposite could be true because it is so popular. I would encourage you to spend some time listening to other podcasts similar to your concept. Is there room for your niche? Are there any voids in

Krakenimages.com/Shutterstock.com

IMAGE 2-2 You should feel somewhat like this, at least on the inside, when you arrive at your podcast topic

other podcasts that yours could specifically address? If so, then there's a good chance you are still on the right path. BTW, having a hard time finding that void? Seems like all possibilities are being discussed and nothing is available? Don't give up, persevere; chances are, you are going to find your spot if you just keep looking.

As we are beginning to find our spot in the podcastmosphere (completely made that word up, don't worry we'll just go with it), we need to run a couple tests to see if it'll work for us. First, depending on which format you are going with, who do you know that could be potential guests? Now, be honest. Can you name five? How about ten? The purpose of this is to begin to get a sense of how far you are going to have to reach in order to find guests. Nothing wrong with thinking to make acquaintances for potential guests; you just need to have an idea of how far you'll have to reach.

Additionally, can you list a dozen potential episode topics based on your concept? For some of us, this is going to be easy. Some of us will speed past a dozen ideas and will approach a couple of dozen ideas, no problem. For those who match that statement, congratulations, you are really on the right path. For those of you who cannot list a dozen, or even half a dozen, don't can the entire idea. Just know, it still needs work and refining. You've still got some work to do before you're ready to move on.

I want to take this opportunity to express how important I feel that it is for you, the podcaster, to not overlook any detail, no matter how small you think it might be. Do not skip steps or move ahead before you're ready. The time you spend now addressing every little aspect of your show is the time you won't be spending in the future fixing problems that could have been prevented. "I cannot teach anybody anything, I can only make them think," Socrates said. I hope you will think about addressing every detail as you go.

Alright, we are moving forward, and we're moving forward with the quotes. "What's in a name? That which we call a rose by any other name would smell as sweet," Juliet said in Shakespeare's *Romeo and Juliet*. This line is delivered by Juliet to Romeo as she explains that a name is nothing but a name, and since a name is merely an understood label, it has no value in and of itself. Now I personally love seventeenth-century comedies, tragedies, and romances. I think it makes for a great topic for podcasts. In fact, I believe some of the very best works ever created in the English language were the writings of Shakespeare. However, I'm going to take the opposite side of William, and I'm going to answer the question, "What's in a name?" Everything.

Daniel Flucke/Shutterstock.com

IMAGE 2-3 Having the proper name for your podcast is a critical step

Your show can get lost, no wait, your show can get buried in a sea of podcasts with a name that sounds flat and creates no excitement. You have to be clear and concise with your podcast name. Don't be vague, be specific. Be creative, use a play on words, grab listeners' attention from the beginning. We talked earlier about a project I am working on in my spare time. Remember the name? I hope you said yes. The vodcast will be titled *Let's Go Smoke*. The name is a play on words; we are talking about all things BBQ and smoking all types of meats. Of course, when one hears the title something else comes to mind and that's the point. Hopefully the name sticks, and then the content delivers, and the viewers come back for more.

> *It is right here where I need to be transparent about this upcoming project. Remember just a few pages ago when I stated most all podcasts follow one of the six formats? Well, that is true. However, for* Let's Go Smoke, *I wanted to do something a little different, and the result was a hybrid of sorts. This upcoming vodcast will be a mixture of interview style and educational style. I'll interview pitmasters and aficionados alike. However, I plan on having entire episodes dedicated to informing viewers of tips, tricks, and methods to improve their BBQ. So, I'll use this point in the book to state that you are more than welcome, if not encouraged, to choose a hybrid style for your podcast.*

I hope the foregoing paragraph didn't introduce confusion; that certainly wasn't the idea. It was simply meant to provide one more format that could be useful in the creating process of your podcast.

Moving forward, the next item on the list of creation is your podcast's logo. Everything up to this point and going forward from this point is an extension of your brand. Therefore, making sure the look and the feel of your logo is exactly what you want your viewers/listeners to see first. Think for just a moment about a professional sports team. Any team will do. Not a sports fan? Not a problem. Think about any store or brand of product you regularly purchase or visit. I'll pick a couple. Starbucks, the Green Bay Packers, Domino's Pizza. Did you see one or more of those logos in your mind? There's a good chance you did, and you know why? Because a great deal of design, thought, and theory went into each one of those for the simple purpose of sticking in your mind. Not one aspect was overlooked, not the font choice, the choice of shapes, or the colors used. Every aspect has a purpose and you should also consider all such aspects while designing your podcast logo.

Don't think you have what it takes to create your own brand? Consider this. In 1971, a graphic design student was contacted to design a logo for a company called Blue Ribbon Sports. The student's name was Carolyn Davidson. She accepted the proposal and was paid $35 dollars to create the company's logo. After some time she produced her work to the company, and her design was accepted. That company changed their name a few years later to stand out in a crowded field of sporting goods. They renamed themselves to Nike and have continued to use the same logo over 50 years now. Understanding this bit of information gives me all the confidence in the world that you can begin creating your brand, your image, and the feel of your podcast. Let's start with some of the basics of logo design.

First, what is the feel of your podcast? Are you going for religion? Perhaps spooky, unsolved murder mystery type. Are you going with positive, encouraging, uplifting? No matter your direction, you need to be thinking about concepts. In fact, the concept is king. A sidestory: back some years ago, NBC wanted to convey their excitement about new color TV. Their concept was to remain excited about color TV. Thus, the colorful peacock was born and is still in use today. Concept is king because it limits you and frees you at the same time. Let's take a look at a concept I was working on recently.

IMAGE 2-4 Burchwood logo #1

The concept was BBQ and rustic. The first line of thought was BBQ always requires fire, and most always, fire is orange and black. And with that, the dominant colors were adopted. Next, we had to decide the shape of the logo. An elongated hexagon was chosen because it is a shape we see everywhere no matter if we recognize it or not. The name Burchwood BBQ was already in place and the "Thumbs Up Tasty" slogan came along a little bit later. The weathered background and the reclaimed barnwood background fit nicely because they accomplished two things: First, they make the orange of the logo pop, and second, and most importantly, they work hand in hand with the concept, which was BBQ and rustic.

Then we wanted to take it a step further. The question was could we move around the dominant colors and still make it work. Take a look at some different versions of the logo, and see if it accomplishes the same objective.

IMAGE 2-5 The same logo with different color variations

Notice how there is very little change between each of these. I encourage you to try the same. Move some of your outline colors around. Add and remove background colors.

The Psychology of Logo Design

Of course we want to have the right shapes, complementary colors, and font size, and we'll address each. But first let's discuss what's behind a powerful, memorable logo. Familiarity, loyalty, and self-identification are just some of the reasons found for customers purchasing certain brands over other brands. Recent research is supporting the idea that consumers, or listeners, use these same motivational factors in consuming digital media, or podcasts. Some additional motivating factors in consuming digital media include social identification and emotional association. And the very first thing your listeners are going to see, before they hear 1 minute of your podcast? You guessed it, your logo. How do you want them to feel when they see your logo? To answer that question, let's start with colors and see what they mean to listeners from the beginning.

Carl Jung is most associated with modern color psychology; however, the theory has been around for hundreds of years. In its most basic form, color psychology states different hues have different effects on human behaviors. Additionally, color can influence and even trigger emotions in humans. While it is impossible to state with absolute truth that the following colors trigger certain emotions and responses to all humans, research supports the theory that the following colors have an emotional and intellectual response.

RED—power, excitement, love, anger, speed

YELLOW—happy, competent

GREEN—good taste, environmental friendly, healthy

BLUE—sophisticated, reliable, competent

PINK—sincere

ORANGE—rugged, exciting

BLACK—expensive, fear, sophistication

WHITE—purity, sincerity

PURPLE—warmth, power

Again, these colors are not going to have the same impact on everyone, but as a general theory, most associate those characteristics with those colors.

Now that we have an idea of what color can trigger in our viewers, let's get a bit further in looking at color combinations that can have further influence.

We want to use a combination of colors that are harmonious. There are several different theories of the color spectrum we'll define so as to give you the tools to find the correct colors. The first is monochromatic, and this is the idea of using one color with different hues, or shades within that color's spectrum.

Followtheflow/Shutterstock.com

IMAGE 2-6 Different hues of brown—monochromatic

The next stage in the color theory is complementary colors. This is defined as using two colors that complement each other in harmony. Usually opposites attract, and functioning complementary colors are red and green, orange and blue, yellow and purple. See Image 2-4 to see how orange and black are opposite but are complementary colors.

designium/Shutterstock.com

IMAGE 2-7 Yellow and blue are opposite and complementary colors

The third of the five color schemes we are going to discuss in this book is triad. This is a method of using three colors that are equal in distance from one another on the color wheel, forming a triangle. Additionally, you can move the triangle in any direction and the colors, as they change, are going to remain complementary. Now there isn't a rule written in the book of logo making that states you cannot exceed three colors for your logo; it's just recommended that you don't. See Image 2-5 to see the triad in use.

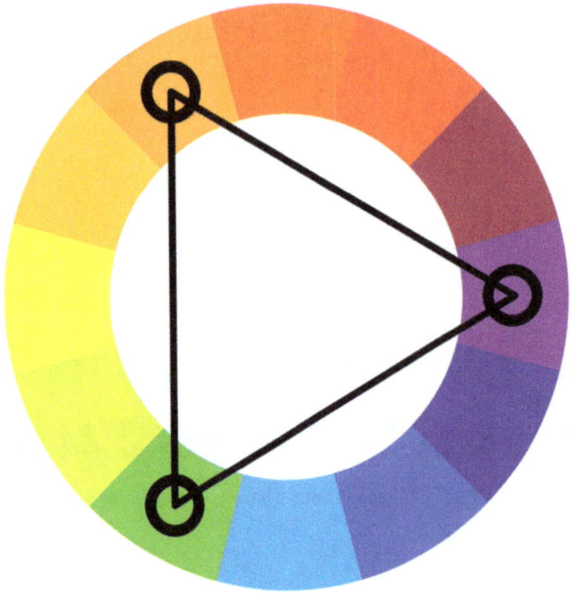
Oleksandr Panasovskyi/Shutterstock.com

IMAGE 2-8 Triadic colors that can move over the entire wheel for three complementary colors

The next color scheme is referred to as the analogous color scheme. This theme essentially uses three hues that are located next to one another on the color wheel. Usually this scheme involves one dominant color, either a primary or secondary color, and then two supporting colors. Image 2-9 shows an example of analogous colors.

Oleksandr Panasovskyi/Shutterstock.com

IMAGE 2-9 Analogous color scheme

Notice how the colors are located next to one another, the dominant color is green, which is a secondary color, and the three colors complement one another.

The final color scheme discussed in this book is the tetradic (double complementary) scheme. This is the most complex of all of the color schemes because it uses four colors that are organized into complementary pairs. To create a logo using this color scheme, the designer needs to have a clear purpose for doing so. Using this color scheme will produce vibrant results but if done wrong can have a negative impact on your branding message. Image 2-10 is an example of the tetradic color scheme.

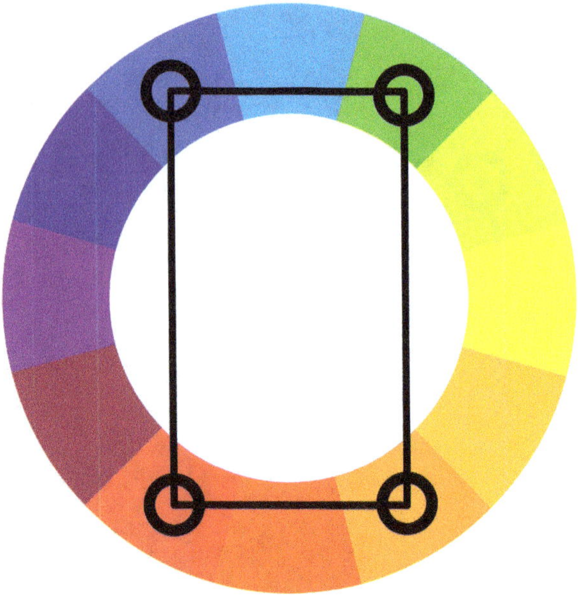

Oleksandr Panasovskyi/Shutterstock.com

IMAGE 2-10 Tetradic color scheme

Remember, concept is king. Create your concept and then begin with your logo creation. As a tip, many sites including Canva and Adobe have free-to-use color wheels that allow you to play with color combinations and color schemes. I'm a firm believer that time spent in this arena is time well spent. You may still feel the urge to jump right in and begin podcasting, I beg you to be patient, we're almost there.

Choosing the Correct Font

As we'll discuss some font types, and what to consider when choosing yours, I just feel like right now is a good time to state, "Logo creation is much more about developing your concept, rather than designing your logo." Simply put, do not rush or overlook the importance of fully developing your why before developing your what. Have purpose for everything within your logo.

So, let's choose a font that's right for you. The following list is a breakdown of some of the more popular font styles and what separates them from other fonts.

Serif fonts are a basic type that has a little character to certain letters. This character is referred to as tags, or sometimes flags. Here is a serif font blown up, with the tags identified.

Podcasting

Times New Roman, Georgia, and Baskerville are some of the most popularly used serif fonts. The use of this font style within your logo reverberates being reliable, being dependable, and having a good reputation. Whole Foods, Mercedes-Benz, and Wikipedia use serif fonts in their logos.

The next type of font to consider is a **sans serif**. The term "sans" means without, and in this case it is without flags or without tags. These are easier to read, and newspapers and online news sources have been using sans serif fonts for years.

Podcasting

No flags or tags, and Arial and Helvetica are a couple of the most common used sans serif fonts. The use of these fonts generally means contemporary, simple, straightforward. Spotify, Jeep, and Google use sans serif fonts.

The next font we'll discuss is script, or cursive, fonts. These are somewhat decorative fonts, but they do have purpose and have been in use for many years.

Podcasting

Script fonts are creative and more lighthearted. They also create the feel of more emotion, more interesting, and even more feminine. Coca-Cola, Kelloggs, and the Los Angeles Dodgers use script font in their branding.

The fourth type of font is called modern font. It is, as the name implies, meant to give a modern, distinctive, forward-thinking image.

PODCASTING

Tesla, Adobe, and Zabka are just a few of the companies that use a modern font for their branding.

The fifth and final font type this book will introduce is novelty fonts, or even referred to as display fonts. These are meant to be fun, wacky, and eccentric.

Podcasting

Nexus and Walt Disney use fonts very similar to novelty or display fonts.

Choosing the Right Shape for Your Logo

As we begin to describe the meanings and feelings associated with using particular shapes and lines in your logo, please remember, there are lots of conflicting theories on this topic. Lots of different theories. The associations made in this book are the ones I've adopted to be true, and to the extent of my knowledge are true. One last point to make, the idea of introducing color, shapes, and font types is for you, the designer, to put together a theme where there is similarity and continuity. Remember your brand, or your logo needs to have a purpose for every aspect of that brand. Themes, feelings, and purpose need to align, otherwise you are creating a brand that could completely miss its intended audience.

OK, with that, let's jump in. In this portion of the chapter we are going to discuss circles, squares, triangle, vertical, horizontal, and abstract shapes.

Let's start with circles. How many logos come to mind when you think of circles? They are, in essence, endless. It is because they represent friendliness, bonding, rapport, and stability. What brand doesn't want to be affiliated with those? This is why circles are the most common shape for logo design. The pros of using a circle is to become identified with the aforementioned characteristics; the cons are it is a busy field: Hard to stand out with so many identifiable brands that use circles. Hard, not impossible, and the circular shape is easily transferable to most of your social media profiles.

Another shape that brings some strong characteristics is the square. Often the square is associated with strength, continuity, long-lasting, and professional. Again, all strong attributes, and when coupled with the right color and fonts, the square can be a powerful shape for your brand.

Next up is the triangle. The brands that choose triangles are usually establishing the image of elegance, sophisticated, educated, and masculinity. This can work quite well in the brand creation of podcasting, assuming it is coupled with the right topic. Important here not to limit your thoughts on just how triangles can be used. Think of the Adidas logo: It is masterfully done in that it is a triangle but made with offsetting lines that tie nicely into the line of sporting apparel they manufacture. This is a good example of no detail, no matter how small one might think, is to be overlooked.

REDPIXEL.PL/Shutterstock.com

IMAGE 2-11 No detail is too small in the logo design process

Let's jump into the logo world of vertical and horizontal line usage in branding, and it is literally as easy as this; vertical lines are associated with sophistication, knowledge, and power, while horizontal lines seek to establish sereneness, peacefulness, and calm. And it really isn't any more complicated than that. Think about how you want your brand to be portrayed and if lines emphasize or enhance use them. If they don't, avoid and don't try to force things.

The final shape is abstract, and perhaps the most fun, that is, when it's used correctly. When it is, abstract branding can allow for standing out, to be recognizable, and to break from tradition. It is popular with start-ups in industries that may not be mainstream. Airbnb is a good example of abstract usage at the time when renting someone's house for a night or two wasn't popular.

So you might be asking yourself, *do shapes really matter in logo design?* The simple and straightforward answer is yes. Very much so. Shapes, along with lines, colors, and font choices, directly influence the way your listeners or viewers are going to interpret how they feel about your brand right from the start. If you take your time, and harmonize the elements of your branding, your audience might just stick around for more than 3 seconds; then you need to deliver on content. More about that down the road.

Where to Start in the World of Software for Creating Logos

I wanted to show you a concept and a few of the rules of logo creation first, before jumping into the world of actually creating your logo. Remember, concept is king. Run freely within the concept, but do not run outside your concept. So, if you have a few, short parameters outlining your concept, let's take a look at some options for logo creation.

If you type logo creation into your preferred search engine, chances are you are going to get way more than you bargained for and you still have no real direction of where to start. There are dozens and dozens of free logo designing software programs available, just keep two things in mind if you are going this direction. First, you truly get what you pay for. Oftentimes, the free generated logos are based on a template that is very vanilla, and oftentimes oversimplified. However, they are not a bad place to start. Here are a few sites that I have used or had students use in the past that are adequate for the beginning of branding creation. Keep in mind some of these free logo generator sites are anything but free. Some allow you to create your logo, but you got to pay to download a high-resolution version. Also, some allow a one-time pay, while most want you to sign up for a service, that is, a monthly fee.

- Hatchful
- DesignEvo
- VistaCreate
- Canva
- Looka
- Tailor Brands
- LogoMakr
- MarkMaker
- Fiverr

These are just a few of the many logo creation and generator sites available. Also, just for clarity, the difference between logo creation sites and logo generator sites is that creation sites allow you to design shapes, lines, text, colors, etc. to your desire, while generator sites get some basic info, name, slogan, industry, and then generate a logo based on some templates. This may work, but as mentioned earlier, they would look vanilla. And since this is your brand, maybe we should be thinking of ways to make it stand out.

When I think about paid-for software, there are two that come to mind. I currently use both due to the high volume of design work we do for clients or for the classroom. The first one is Canva. They were mentioned in the free version, and they do have a free plan. However, for a moderate robust design software, with a simplified user dashboard, coupled with a very affordable monthly price, Canva is a strong choice for design. The second choice, and probably the one I use most, is the Adobe Creative Suite. The complete suite has the ability to do many, many creative projects. If you are pursuing higher education, there are affordable monthly plans depending on how in-depth you need to be with your designing software. User interfaces and tools are similar from program to program, and this allows for some easier migration between the suite. However, the learning curve for most of the Adobe programs is extensive. These are very robust programs that can feel a little overwhelming. These are some of the most complex yet highest quality producing design software programs available. If you are willing to jump in and learn, or you have some experience and want to know more on how these programs work, my opinion is you cannot go wrong with spending the monthly subscription price for the high-end content you can create with Adobe Creative Suite.

OK, we've covered a lot in the pages of this chapter. Let's take as long as we need in order to look over the checklist, ask and answer the questions, and get to know those terms. They are going to be very handy in the coming pages. At this rate, we'll be recording with purpose in no time.

Rawpixel.com/Shutterstock.com

IMAGE 2-12 Logos say a lot about your podcast; work on yours until it's right

Chapter 2 Terms

Analogous color scheme Uses three hues that are located next to one another on the color wheel. Usually, this scheme involves one dominant color, either a primary or secondary color, and then two supporting colors.

Complementary colors Using two colors that complement each other in harmony. Usually opposites attract, and functioning complementary colors are red and green, orange and blue, yellow and purple.

Conversational-style podcast Format where a host and a cohost converse over a variety of topics.

Educational-style podcast Format meant to inform listeners on specific topics.

Interview-style podcast Format where the host and cohost have a variety of guests to discuss certain topics. Topics can vary, though the overarching theme is consistent.

Modern font It is, as the name implies, meant to give a modern, distinctive, forward-thinking image.

Monochromatic The idea of using one color with different hues, or shades within that color's spectrum.

Nonfiction-style podcast Format where the host and cohost cover a wide range of topics that are meant to inform listeners.

Novelty/display fonts These are meant to be fun, wacky, and eccentric.

Panel-style podcast The host and cohost have a different panel of experts to discuss specific topics. Guests are knowledgeable in the field of the discussion.

Sans serif fonts The term "sans" means without, and in this case it is letters without flags or without tags.

Script fonts Cursive fonts; these are somewhat decorative fonts, but they do have purpose and have been in use for many years.

Serif fonts Basic type that has a little character to certain letters. This character is referred to as tags, or flags.

Tetradic (double complementary) scheme The most complex of all of the color schemes because it uses four colors that are organized into complementary pairs.

Theater-style podcast Format is fiction and entertaining for the listeners.

Triad This is a method of using three colors that are equal in distance from one another on the color wheel, forming a triangle.

Chapter 2 Checklist

- I've read Chapter 2 and have a firm understanding of its context.
- I'm completely sure I know why I choose the podcast format that I did.
- I know the colors for my branding.
- I know the font choice for my branding.
- I fully understand why I choose these colors.
- This is the shape of my branding.
- These are associations I wish my listeners/viewers to make with my branding.
- Everything element in my logo has purpose.

Chapter 2 Questions to Ask Yourself

(Be honest with yourself, you might as well be.)

1. Do I have potentially ten guests or episode topics for my podcast?
2. Do I fully understand my purpose for why I am getting into podcasting?
3. Do I know why I choose each of my logo/branding elements?
4. Can I answer the concept for my branding in one, simple sentence?
5. Do I fully understand the reason why I've chosen this podcast format?

3

Setting Up Your Studio

Alright, now we are beginning to get to the good stuff. We've swam through the importance of understanding your branding, your podcast format, and preparing for the journey ahead. Hopefully we haven't forgotten we need a plan, purpose, and perseverance. As you begin to set up your studio, let me encourage you to not forget these p's. In fact, they just might appear in the Chapter 3 Terms section at the end of this chapter. However, let's get back to the point at hand, and that is setting up your studio. In this chapter, we'll break down how to prepare for the simplest, most cost-effective studios to elaborate, full-production studios. I am grateful you fully understand the format you'll be using for your studio, because without that, how could we know how to design your studio?

What You'll Need

No matter how elaborate your studio is, there are some basic pieces you are going to need. We'll make a list of equipment, and then begin to dive in deeper to explain the function of each piece of equipment, and the choices you'll need to make in order to get the right equipment for your studio.

- Microphones and mic stands
- Proper cabling
- Headphones
- Audio board
- Means to record
- Soundproofing

This should get you started with a very basic studio. In fact, we've probably all heard of beginning your podcast in your closet. Well, the truth is you absolutely could. Most closets are small, interior (allowing for less outside noises, and good acoustics), and most closets are full of clothes which absorbs sound waves and prevents those waves from bouncing off walls, creating echoes. But we are going to take one step up from there and start with a small, empty room.

Now that we have a space in mind, let's begin with microphones and which one is right for you. In its simplest form, microphones work as transducers and they convert sound waves into audio signals. There are three different types of microphones we are going to discuss in this book, namely, dynamic, condenser, and ribbon microphones. Let's start with the dynamic microphone and learn how it works.

As you speak, you are creating sound waves, and these waves of energy are directed toward the microphone. It is important to note that all the sounds you hear are vibrations being carried in the air around you. When these sound waves enter into a dynamic microphone, a diaphragm inside the mic moves back and forth with the rhythm of the waves being spoken into the mic. A coil is attached to the diaphragm and this coil also moves with the diaphragm. The coil is wrapped around a magnet, and as the coil moves it creates a magnetic field. As the coil moves through this magnetic field, electric current flows through the coil. The electric current moves through the microphone to an amplifier or to a recorder, and just like that, the microphone has converted your sound waves into electrical energy.

Condenser microphones work very similar to a capacitor. When you speak into a condenser mic, your sound waves hit a diaphragm. In the condenser mic, the diaphragm is actually one of two metal plates. As the diaphragm vibrates, the distance between the two metal plates shortens and expands. It is during this movement that charge is created and the condenser mic converts this charge into electric current and moves it out of the mic to an amplifier or to a recorder. There are large diaphragm mics and small diaphragm mics. Simply put, the small diaphragm mics are a good universal mic, and a large diaphragm mic is good for creating deeper, more professional audio.

Ribbon microphones work by using a thin vibrating ribbon of metals, thinner than the plates in the condenser mics, that move between the poles of a magnet. These mics do not use diaphragms and coils; instead, the movement between the poles generates the electric current, which moves through the microphone to an amplifier or to a recorder.

There are numerous variations of these microphones, but you need the one that is best for you. How do you decide? When you are deciding on the correct microphone for you, it is important to understand the mic's polar pattern, or the direction from which the mic picks up sound. Some mics pick up sound from the front and others from the front and from the sides.

There are three distinct options when considering microphone polar patterns. These options are cardioid mics, omni mics, and bidirectional mics.

Let's start with the cardioid mics. These mics are unidirectional and pick up much better sound from the front of the mic, rather than the sides. The omni mic is great for picking up all sounds around you, great for natural sound. And the bidirectional mic is best for picking up sound from the sides of the mic, and not so much from the front.

> ✻ When you see this fancy asterisk, that is when you know I'm about to share my opinion. Again, it is only my opinion and I encourage you to research for yourself. However, for podcasting I think a large diaphragm condenser is a solid choice, and one you won't regret. Currently, I am using Shure SM7B mics and love the results.

darksoul72/Shutterstock.com

IMAGE 3-1 Shure SM7B is a good choice for a podcasting mic

When thinking about your mic stand, most of your mic stands are going to have one of two basic designs, desktop mic stands and boom-scissor style mic stands. Image 3-2 shows a variety of desktop mic stands.

New Africa/Shutterstock.com

IMAGE 3-2 There are a variety of mics stands available. Take the time to consider which one is right for you

Desktop stands come in different styles, one option to be aware of is the ability to adjust height in your mic stand. Every guest is going to be a different size, be adjustable.

Image 3-3 is your boom-scissor style mic stand—very versatile, and easily adjustable to different heights of your guests.

DC Studio/Shutterstock.com

IMAGE 3-3 Boom-scissor style mic stand

You can clamp this type of mic to the desktop, and it is easily adjustable. I've used both types of mic stands and both are good. Currently, we are using desktop stands and the only drawback is occasionally you will have guests where the mic is almost in the way of the shot. Easy enough to work around—however, the next studio will have the clamp-scissor style mic stand.

Getting the Proper Cables

Depending on which type of microphone you choose, you are going to need the proper cabling to use the mics you choose. The most popular connector for professional mics is the three-pin XLR connector. Image 3-4 is your standard XLR three-pin cable.

Max Coen/Shutterstock.com

IMAGE 3-4 XLR Cable

This is a standard XLR, male-to-female, three-pin connector; pin 1 is ground, pin 2 is positive polarity mic audio signal, and pin 3 is negative polarity mic audio signal. The SHURE SM7B is XLR connected.

Before we go all in on cabling, there is a matter we need to address, and that is balanced versus unbalanced audio. Since I don't want to fill all of these pages with the pros and cons of balanced versus unbalanced, and get myself lost in the process, I'm going to use Kelly Johnson's KISS advice and that is "Keep It Simple Stupid." In so many words, an unbalanced audio cable is carrying an audio signal using two lines: a signal and a ground. The ground wire is simply a reference point for the audio wire, though the ground wire can pick up noise along the way, resulting in buzzing or humming in your recording. Where does that noise come from? Well, if you think of your ground wire in an unbalanced cable like an antenna, it can pick up noise from just about anywhere, most commonly (if your setup looks anything like mine) you've got wire spaghetti behind your equipment. When wires cross one another, unbalanced lines can pick up noise from simply lying across other cables. Images 3-5 and 3-6 are examples of unbalanced cables.

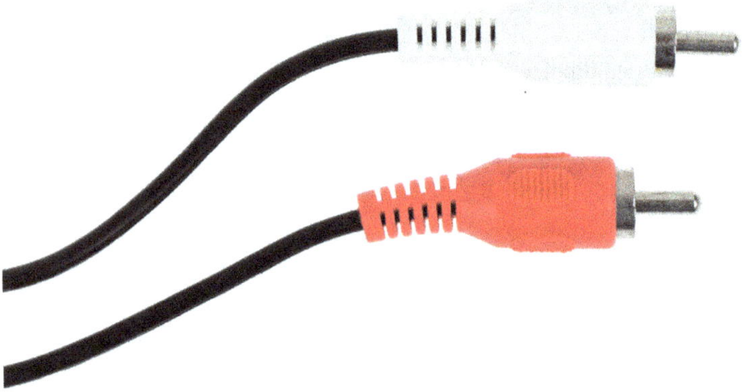

MyImages - Micha/Shutterstock.com

IMAGE 3-5 RCA jack

RCA jacks are not as popular as they once were, but they are unbalanced audio cables.

Fernati2007/Shutterstock.com

IMAGE 3-6 Quarter-inch TS cable

The quarter-inch TS cable is unbalanced; think connecting a guitar to an amplifier. The tip is the signal and the sleeve is the ground.

Then you have balanced audio. Remember when we discussed the three-pin XLR, pin 1 is ground, pin 2 is positive polarity mic audio signal, and pin 3 is negative polarity mic audio signal? This is the difference between balanced and unbalanced audio cables. Balanced cables have a second audio signal line; the positive line is "hot" and the negative line is "cold." When arriving at whatever audio board, or recording device the balanced cable is plugged into, the "hot" and "cold" wires signal is flipped, canceling out any noise picked up along the way. Of course, we know the XLR is balanced, Image 3-7 shows a tip–ring–sleeve (TRS) cable.

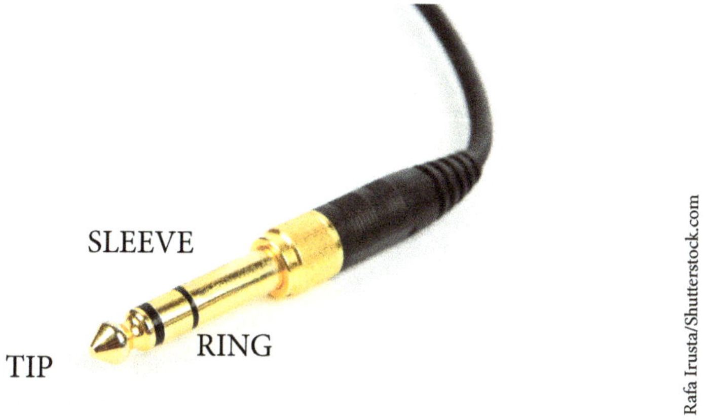

IMAGE 3-7 TRS cable

Notice the difference between the TS cable and the TRS cable? The TRS has a second ring, and this indicates a balanced cable. The tip is the "hot" or positive signal, the ring is the "cold" or negative signal, and the sleeve is the ground. This accomplishes the same purpose as an XLR cable and just uses different inputs for different pieces of equipment.

So, which one is right for you? Table 3-1 lists some pros and cons for each type of audio cable.

TABLE 3-1 Pros and cons of mentioned cable types

TYPE	PROS	CONS
TS—unbalanced cables	- If you are setting up a small studio, and the cable run is short, say, 10 feet or less, TS cables should stay relatively quiet. - Least expensive of any of the mentioned cables	- Since the TS cable has only one signal carrying line, TS cables cannot carry stereo signals. - There is no way to assure you won't pick up noise along the way.
TRS—balanced cable	- Most often doesn't require a preamp (more on that in a moment) - Both ends of the cable are male, so connections are simple. - In most cases, TRS is less expensive and requires less additional pieces of equipment	- Bottom line, our production is about audio quality, and the TRS isn't the gold standard for delivering professional audio.
XLR—balanced cable	- Always balanced signal - Based on the design, the cables snap into most inputs, making it more difficult to accidentally become unplugged. - With a preamp, XLR delivers the highest quality audio in a studio setting.	- The most expensive cable of the three listed - In some cases, XLR requires a preamp in connection with condensed microphone, thus incurring more costs to get started.

Before I throw in my two cents, let's discuss that preamp. Oftentimes, depending on the audio board or switcher you choose, your XLR mics are going to need a preamp before plugging them into your board. The preamp accomplishes two things: First, it provides the correct load impedance to match the microphones output, and second, it boosts the signal from the microphone to a usable level. Image 3-8 is a preamp.

IMAGE 3-8 Cloudlifter preamp for XLR microphones

Cloudlifter is one of many brands of XLR preamps. You plug your XLR cable into the preamp from the microphone, and then with a second set of XLR cables, plug into the opposite side of the preamp, and then into your audio board, and BAM! Your audio is amped and usable.

> * Remember this little asterisk from a few pages earlier? That's right; it's opinion time. For the very best in podcasting audio, connect your SHURE SM7B mics via XLR cables to a Cloudlifter preamp and into your audio board for the best, clean, noise-free recording you can get.

Before we jump into audio boards, let's take a brief look at headphones. Because while all headphones accomplish the same thing, it's important to understand some differences between makes, and since you're going to be wearing them—a lot—let's get the ones that are right for you.

Headphones: Eeny Meeny Miny Moe

How to pick out the proper set of headphones? Well, several elements need to be considered. *Are they for studio use only? Is there a pair that I can use for stu-*

dio and for editing? How much do I have to spend on a quality set of headphones? All are good questions that you could ask yourself and all need addressing before you buy your next set.

sergey causelove/Shutterstock.com

IMAGE 3-9 There are many different styles and choices of headphones—get the one that's right for you

There are different types of headphones. Let's discuss open back versus closed back. Open back headphones are simply as they sound. They have openings to the back of the headphones to allow for air to pass. These style headphones have a very natural sound and allow the listener to hear everything going on around them. But that in itself is a problem. When you are in the studio, you do not always want to be able to hear all sounds around you but rather just the sounds coming from the microphones. Also, the open back headphones will have sound leakage, meaning sounds coming through these headphones will no doubt spill out into the studio where hot mics will pick up that sound. These headphones have a place and a purpose. Gamers are big on open back headphones for a variety of reasons; however, they probably should stay out of the recording studio.

Your closed back headphone has essentially four elements to consider before you purchase. The first is isolation, and as it states, you are going to want a pair that isolates noise from leaking out into the studio, as well as keeping intended sound inside the headphone. The second determining factor should be minimal coloration or sonic neutrality. These terms simply refer to the headphones being able to reveal any defects or faults in the recording. Being able to identify these preproduction will save time and sanity in the editing bays. Comfort is another factor that should be considered. If you have the opportunity to try on a pair of headphones before you buy them, do it. Everyone's head is different, and there is no umbrella answer to the best comfort. A pair with lots of padding is a good recommendation—again you'll be wearing these a lot. Finally, the last thing to consider is price. You can spend as much as you want to on headphones. You can spend over $5,000 for a pair of studio headphones; don't do this. You can spend as little as $35 for a pair of studio headphones; don't do this either.

✱ If you want a good entry price, coupled with quality sound, and comfortable wearing for long periods of time, I suggest looking into the Audio-Technica ATH-M series. You'll find numerous options at a variety of prices. I've used several models of these for years, and based on your studio budget, you won't be sorry with any of these options. They offer a wide range of headphones for your recording and for your editing bays (more on this in Chapter 6).

Audio Boards: Now We're Getting to the Fun Stuff

OK, let's slow for just a second. Before we plunge into the world of options of audio boards that are available, let's first get a good understanding of what the board does, and how that relates specifically to your podcast. To clarify, I call it an audio board; as you begin your search for the board that is right for you, it's likely you are going to hear the board referred to as a mixer, a sound board, a mixing console, or a mixing board. Don't get confused; they are in most part the same thing, and they serve the same purpose. And that purpose is to take multiple input signals from peripherals such as microphones and merge them together to create one signal that can be sent to your recording device for later use. That, in a nutshell, is it. Now, let me not understate the enormous varieties that exist in audio boards. There are very simple two-input boards to very complex 72-channel, 24-bus consoles with 48 bus split and linked bus mode modifications. We'll probably be using something in-between.

Image 3-9 shows a RØDECaster Pro IIaudio board. This is a good, entry-level audio board that is capable of producing good audio for your podcast, as well as being versatile for upgrades you'd make to your studio.

IMAGE 3-9 RØDECaster Pro II audio board

Easy to use, and easy to adjust. This is a four-channel audio board with 16 on-board sounds for your podcast (think audience applause, or audience laughter). You can add or delete sounds to fit your show. There is a touchscreen for easy setting adjustments, and recording is as easy as hitting the record button. Image 3-10 is the back of this unit. Let's take a look and see if we can recognize any of the inputs on this board.

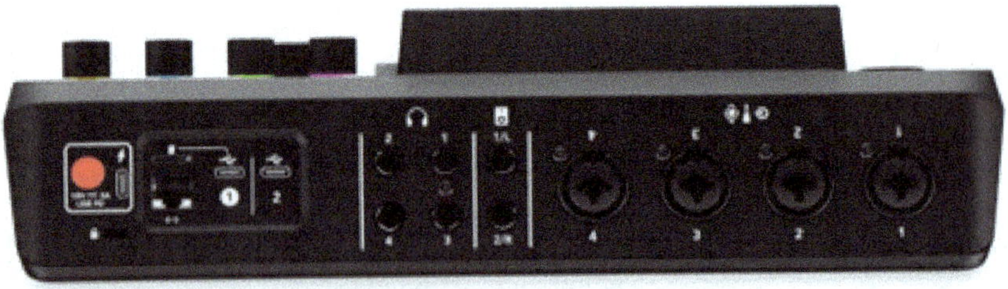

IMAGE 3-10 The back of the RØDECaster Pro II

As you will notice, this board has four XLR inputs and four corresponding headphone jacks. I'm sure you've done the math and figured out you can have up to four mics and headphones into this board, which would provide you the opportunity to have up to four live persons on any one episode of your podcast. Additionally, this board has two line-ins, a right and a left, for stereo, iPhone/Android input capabilities for using prerecorded audio, and a microSD card slot for recording directly from the board.

Again, I am in no way promoting any brands or manufacturers of studio equipment, and I believe it's equally important to understand that I'm not saying these described products are the best, the cheapest, or the best built. I'm simply sharing what I use and the equipment I am familiar with. I know there are many, many options and I beg of you to explore the choices that best fit your needs when designing your studio.

That said, the RØDECaster Pro II is a great start for a podcast studio. The audio is clean and crisp. But let's say you had wanted to take your studio up a notch, and had wanted to get into vodcasting—well, stick around. In the second half of this chapter, we'll discuss switchers, lightning, studio sets, and design in order to create your full-production studio.

For right now, let's do a quick recap for our audio podcast studio. To this point we have our microphones, our cabling, and our preamps, we have our headphones, and now we have our audio board with a means to record. In other words, we only lack soundproofing to get this thing off the ground.

Soundproofing: Is This Part Necessary?

In so many words, yes. We want the cleanest audio possible and in order to prevent creating work for ourselves, we want the cleanest audio possible on the first recording. Soundproofing your studio will give you a much higher quality product. The purpose of soundproofing is to keep your recordings in its space—to prevent sound from leaving the room, or to prevent sound from entering the room. Sound-absorbing panels will soak up extra sounds that can create echoes, or background noise.

While there are multiple soundproofing options, from sound curtains, to blown-in soundproofing wall insulation, to soundproof drywall, there are two options that are practical, affordable, and moveable (you know when you're going to need a bigger studio). The first of these are simple acoustic foam tiles. These are usually 12 × 12, are available in self-adhesive options, and do a remarkable job. Image 3-11 is a wall of acoustic foam tiles.

SA Production/Shutterstock.com

IMAGE 3-11 Wall acoustic panels

When alternating the layout of these tiles, just as in Image 3-11, acoustic foam acts as a nice absorber of sound and greatly reduces echo. I built a studio using these exact tiles and I honestly have no complaints. The pros are they are affordable and easy to install. I used a spray adhesive and they're still standing years later. The only cons to using these are, depending on the size of your studio, you're going to need a lot of these. Remember, these are 1 square foot each, and if you are soundproofing from ceiling to floor, you really need to measure to get a total wall square footage before ordering (I know this from experience; you see I didn't do that and had to place two orders, and pay twice for shipping. Please avoid doing this).

The second type of soundproofing I'd recommend is acoustic panels. These are a little more expensive, however you'll notice the difference in audio the second you take them out of the box, before they are even hung on the walls. Image 3-12 shows acoustic panels.

chattap pangtha/Shutterstock.com

IMAGE 3-12 Acoustic panels

These panels come in many different sizes and colors. A major bonus to these panels is the ability to spread them out. You do not need to cover all your wall space. My current studio uses 2 ft × 4 ft wall panels, spaced a foot apart, on all four walls and the acoustics and sound absorption are very high quality. I also use these in audio and editing bays and honestly have zero cons.

So, Now What?

We have reached an important mile marker in our journey to podcasting. We are currently at the crossroads of getting started with scripts, show rundowns, and guests (Chapter 4) and moving on to the second half of this chapter to discuss full-production studios. Whichever way you go, don't forget to stop by and spend some time at the end of this chapter. Terms, checklists, and questions are awaiting your arrival.

Full-Production Studios: It's Time to Get Serious

OK, I see you've decided to continue on with this journey of studio building. Since you're still here, I'll share a little secret about this half of the chapter.

This half is intended for ultraserious podcasters—the ones who see themselves operating a full-production studio and seeing the long-term picture. Before we dive into the next step in building your studio, let's get familiar with a few terms we'll be using so that we understand individual roles when we discuss those in just a little bit.

Disclaimer: The exact definitions of these terms do vary. I'm not claiming to be the final word in using particular terms with their definitions. By this I mean you may have used slightly different terms for a role within a studio, and that is perfectly fine. These are the terms, roles, and definitions I know.

In your full-production studio it helps if you have a small crew working together to help the production. If not, that's OK as well. If so, you are going to need a director. This is the person who is in charge of the final product. They make decisions on just about every aspect of the show. Speaking of shows, since we are now producing a vodcast, you are going to need talent. This refers to the individual/individuals who are in front of the camera, who are hosting the show. Every good production is going to need a technical director, and this is the person who runs the audio board, the switcher, and operates graphics. Finally, you may or may not, depending on the size of your production, need a floor director. This is the individual who monitors camera shots, lighting on talent to avoid shadows, and works directly with the director. This role is typically reserved for larger, television studio productions, and is optional for a smaller operation.

Alright, let's jump in. The main difference between a podcast and a vodcast is full audio and video. So, the first thing we need to explore is which camera is right for you. The good thing about studio equipment today is that you can get as elaborate as you wish, or you could answer a couple of questions and begin this leg of the journey getting just what you need to accomplish your objective. Ask yourself, *What am I wanting to do with my vodcast? Stream only? Record for editing later? And perhaps most important, what is my budget?*

Once you have the answers to those questions, let's discuss how you set up your shots. Most, but not all, multicamera studio setups have three cameras. The purpose of this is to be able to switch views between tight shots on talent left and talent right. The middle camera is to be able to capture a wide shot of the entire set. Image 3-13 shows a simple three camera setup with talent, and camera direction.

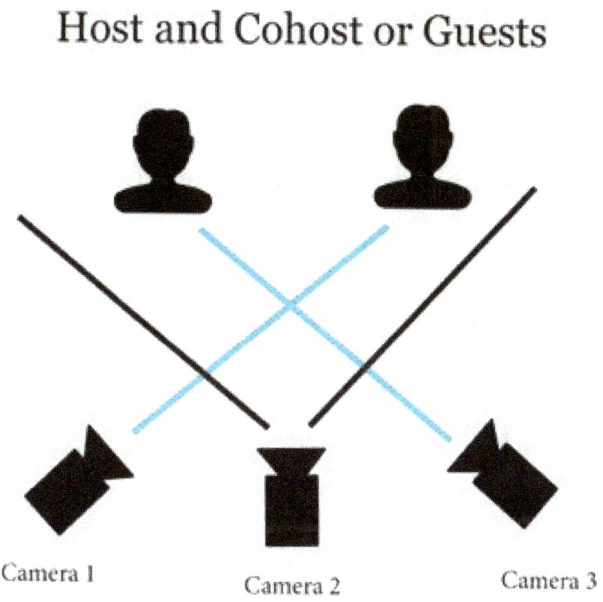

IMAGE 3-13 A multicamera setup.

This isn't as confusing as it looks. Camera 1 has a tight shot unto cohost or guest. Camera 2 has a wide shot on the entire set, and camera 3 has a tight shot on the host. These three cameras are connected to the switcher and operated by the technical director (more on that in just a little while).

When thinking about cameras, we know we are going to need three of them. Cameras today can do many, many things. I find that, as of this publication, that if you stick to a limited number of features, you can get your studio equipped for a reasonable cost. You need to be able to record digital video in at least HD for streaming; 4K is nice but not mandatory for most podcast studios. Your camera needs to be able to record to SD cards and have a 20× optical zoom, SDI outputs, dynamic image stabilization, just in case (which is nice), and a touchscreen view panel. Many manufacturers make camcorders that are capable of fulfilling all these requirements. There is a large price range, so time spent researching is worth it.

Briefly, since I just introduced SDI, let me explain in a little more detail. SDI is short for "serial digital interface." Basically, using a coaxial-like cable, this interface allows for the transmission of uncompressed video signals. The alternative is HDMI, and while both interfaces will accomplish the same objective, SDI is found on higher-level, professional equipment, HDMI is more entry level to prosumer level.

> * If you want a nice, reliable set of cameras that are going record high-quality video, and cameras that are easy to set up and easy to use, consider the Canon XA 15 Camcorder. I've been using this for a couple of years and so far, so good. No cons that I can list.

Next you're going to need to decide what those cameras are mounted to. Are you going to mount on tripods? If so, let me suggest looking for tripods that are heavy duty, with wheels that lock, and have a wide range of adjustable heights. Your camera height is approximately the same height as your talent, but you are going to want the flexibility at some point. Image 3-14 is a tripod with lights.

A heavy-duty tripod with wheels and locks is going to have the ability to hold a wide range of lights. You are going to want versatility and sturdiness. Ideally, your lights will move very little; however, no two sets of talent with guests are the same, and having options to eliminate shadows or poorly lit faces is a plus.

Nikola Spasenoski/Shutterstock.com

IMAGE 3-14 A heavy-duty tripod

Another option for your lighting, as well as your cameras, is a stationary mounting system. A lighting truss system can be ideal for permanently mounting cameras and lights. I know your first thought might be that lighting trusses are for outdoor concerts and bigtime productions. While this is true, truss manufacturers have seen the demand for indoor, smaller systems, and make a variety of models for podcast studios. Image 3-15 is a lighting truss meant for mounting cameras, lights, and even monitors.

FOTOGRIN/Shutterstock.com

IMAGE 3-15 Lighting truss

There are all sorts of mounting accessories available for models like the one shown in Image 3-15. The pros with using a system like this is equipment doesn't move; cords for plugging in lights, and cameras can also be mounted to the truss, making for a cleaner, less hazardous working space. A couple of cons are they can get pricey, they require bigger spaces, and they can get tall. In fact, my last studio build got put on hold due to the truss exceeding fire code height, so an alternate plan was made and I went with tripods. Fortunately I've still got the truss for our next studio.

Speaking of lighting, what kind of lights do we need? Good question. Studio lights have come a long way over the years. No more of the old school, harsh lighting, way too hot lights. Today, there is an endless choice of LED studio lighting sets available. First thing to consider is how much space you need to light, this will help determine how many panels you need. Most kits come with two or three panels. I get one or two extra panels with each build, just in case. Your lights should have 3200–5600K CCT, and RGBW control for color and brightness. They should also be around 10 in long, 9 in tall and 2–3 inches wide. If you can get these with Bluetooth abilities you'll save yourself some headaches.

✻ For the price and the performance, I would recommend a GVM 1200D RGB LED light panel. These come in three light kits and are reliable, bright, and long-lasting. The stands that come with the kit are fine, although I opted for something more sturdy, with wheels and locks.

We've got our cameras, lights, and tripods or trusses; now we need a switcher, backdrop, stage lighting, monitors, computer for digital interface and recording, and furniture. This sounds like a lot; however, just stick with me and we'll get a starting point for everything. Image 3-16 is a look at our recently completed podcast studio on the campus of Tennessee Tech University.

IMAGE 3-16 Podcast studio at Tennessee Tech University

Let's begin with stage lighting. Do you see the blue lights shining up from the floor? These are basic DJ lights with LED bulbs that are hard wired with about 60 different color options to create the feel of your vodcast. These are available on Amazon, and through B&H Photo/Video. There are many different manufacturers; I've tried several and basically can say they are all pretty much the same. They provide good, bright lighting, have easy controls for adjusting color of light, and are easily breakable if not secured and out of the path where people will be walking.

Next is the monitor. This is an HD, smart TV, with multiple USB and HDMI inputs. The purpose of the monitor is for viewers to see graphics that accompany the vodcast. In the case of Image 3-16, what we see on the monitor is the Tennessee Tech official logo and we use this for our local sport show. Fortunately TV monitors have come way down in price. This is a 60 in monitor and fits perfectly for our studio. We create our graphics and save them to a USB flash drive. Input the drive in the TV, select source, click on your USB, select the correct graphic for the show, and your set takes on a custom feel designed for your particular show.

The next item to discuss is the backdrop, and this is back of the set. There are many different ways to create your backdrop. In this case, we built a standard wall with studs separated at 16 in. I bought reconditioned barnwood to create the feel we were going for. We cut each piece to fit in a staggered pattern, and the result is a different feel, when changing floor light color, for every show.

Set furniture is just plain tricky. I say that because I thought it would be easy to order a studio news desk, in my mind I thought there existed such a place where news desks were lined up like new Chevy trucks ready to be sold. What I found out was the complete opposite. There doesn't exist such a store, and my research resulted in one recurring theme: news desks are custom made with a very hefty custom price. Since this wasn't an option I did the next best thing. In my town there are four custom woodworkers who make fine furniture. I met with several of them, discussed exactly what we were looking for, got great pricing, and within 6 weeks had a beautiful, custom desk at less than half the prices quoted from national news desk custom manufacturers. by the way, those chairs came from university surplus. They are Forever chairs but they are working great for right now.

Let's move now to what I believe to be the most important piece of equipment in the studio, and that is the switcher. In Image 3-16, the switcher is the main focal point in the foreground. The switcher is where the magic happens. Your studio cameras are hooked up into the switcher, allowing for the technical director to switch between camera angles during live recording. Additionally, you can have graphics stored on the switchers so that the technical engineer can switch between lower thirds and over-the-shoulder graphics during live recording, saving your time in the edit bay. The lower third is a graphic that appears in the lower third of the screen usually containing host/cohost/guests name and perhaps a title or some other identifying information. The over-the-shoulder graphic is a graphic that appears over the shoulder of the host during the live recording.

There exists an entire universe of switchers that can easily overwhelm you in the beginning. Let's keep it simple and be specific to what you need your switcher to do. In our case, we have a three SDI camera setup; this is rather simple. Switchers can have as few as two camera inputs to as many as 10. No sense in spending the extra money on a switcher that is capable of very complex recording if you are only using several inputs. Blackmagic, JVC, Roland, and NewTek are just a few of the manufacturers of video switchers. A couple of features you need to look out for when considering which switcher is right for you are:

- Build
- Inputs/outputs
- Special effects
- Ability to record to the board
- Price

Let's discuss the build. Switchers typically come in one of two designs. The first is a format where the switcher is mounted in a rack with other equipment and connected to a control surface on the desk of the technical director. This is a common method, although these systems are complex and designed for complex television studios. The second format is a switcher that is integrated into one piece of equipment, such as the switcher in Image 3-16. These are plenty capable of handling the needs of a smaller production, and are more user friendly.

For the input/output section, you'll want to make sure you choose compatible camera outputs to switcher inputs. No matter if you are using SDI or HDMI, make sure your switcher is capable. Also, your switcher is going to need audio inputs from your audio board. Image 3-17 is the back of the board I'm currently using, with some detail on inputs.

IMAGE 3-17 Rear view of an SDI multichannel video switcher

This unit has multiple SDI inputs, headphone input, and two XLR audio inputs. I have found the best audio results come from running mics into Cloudlifter preamps, and then into the RØDECaster Pro II. From the audio board, run line out TPS to XLR cables from the audio board to the switcher.

From the SDI output program, I run SDI cables to a device known as a Spark Plus; this is a NewTek device used for connecting to your monitor so that you can visualize what is in preview and program.

Special effects and the ability to store graphics on the switcher are something to consider. It's just as easy to apply these elements in postproduction; however, if done during the live recording you'll save yourself time. Video transitions from camera shot to camera shot are swipes, or cross dissolves. If this is important to your production, and you choose not to add these post production, make sure your board has these features built in. The same with graphics, if you wish to add these during live recording, make sure your board can store these graphics to be used later. Otherwise, we'll add these during editing (Chapter 6).

Not all switchers are created equally; in some you can record your shows directly to the switchers, while in others you cannot. If your switcher allows you to record to the board, excellent! If not, you are going to need a computer to record to. Either setup can accomplish your objective; however, if you record to your computer, try to record to the computer with your editing software, it'll make transfer life a lot easier.

> * Of course, price may be the most important factor when determining which video switcher is right for you. As you start to research, you'll find prices have an enormous range. If you are looking for an advanced video switcher with a surface board that is capable of handling complex studio setups, the NewTek TriCaster TC1 switcher is a great place to start. At the time of writing, that setup starts around $23,000. For the three-camera, smaller production, Blackmagic makes the ATEM Mini Pro with HDMI inputs for less than $700, and the ATEM Mini SDI Pro for your SDI camera connections for just under $800. My next studio will have the ATEM Mini SDI Pro; it's compact, robust, and mobile if you should ever need to be on location to record.

A Brief Rewind

OK, we've come a long way. Hopefully you are getting an idea of what your studio is going to look and feel like. I bring up "feel" because you want your hosts and guests to feel comfortable so that natural, organic conversation can occur. Take the necessary time to plan your studio. As I previously emphasized the importance of taking time to design your brand, I also suggest you take time to design your studio. It plays an important role in the success of your studio. In the meantime, let's take this opportunity to go over terms, questions, and checklists for this chapter.

Chapter 3 Terms

Audio board A board capable of taking multiple input signals such as microphones, and merge them together to create one signal that can be sent to your recording device for later use.

Balanced audio An audio cable that contains three wires, one positive audio signal, one negative audio signal, and one ground.

Condenser microphone When you speak into a condenser mic, your sound waves hit a diaphragm. In the condenser mic, the diaphragm is actually one of two metal plates. As the diaphragm vibrates, the distance between these two metal plates shortens and expands. It is during this movement that charge is created and the condenser mic converts this charge into electric current and moves it out of the mic to an amplifier or to a recorder.

Director This is the person who is in charge of the final product. They make decisions on just about every aspect of the show.

Dynamic microphone Designed with a diaphragm that absorbs and moves with incoming sound waves. An attached coil also moves around a magnet creating a magnetic field. As the coil moves through this magnetic field, electric current flows through the coil. The electric current moves through the microphone to an amplifier or to a recorder.

Floor director The person who monitors camera shots, lighting on talent to avoid shadows, and works directly with the director.

Ribbon microphone Works by using a thin vibrating ribbon of metals, thinner than the plates in the condenser mics, that move between the poles of a magnet. These mics do not use diaphragms and coils; instead, the movement between the poles generates the electric current, which moves through the microphone to an amplifier or to a recorder.

Technical director The person who runs the audio board, the switcher, and operates graphics during live recording.

TRS cable A balanced cable consisting of a tip, a ring, and a sleeve.

TS cable An unbalanced cable consisting of a tip and a sleeve

Unbalanced audio An audio cable that contains two wires, one is the signal and the other is the ground.

Video switcher Your studio cameras are hooked up into the switcher, allowing for the technical director to switch between camera angles during live recording.

XLR three-pin cable A standard audio cable, male-to-female connector. Pin 1 is ground, pin 2 is positive polarity mic audio signal, and pin 3 is negative polarity mic audio signal.

Chapter 3 Checklist

- I've read Chapter 3 and have a firm understanding of its context.
- I have a good understanding of what is required to put together a studio.
- Before I began planning my studio, I know what my budget is.
- I have researched, for myself, the equipment needed to begin building my studio.
- I'm designing my studio to be capable of changing as my show changes.
- I have a reasonable timeframe for completion of my studio.

Chapter 3 Questions to Ask Yourself

1. Is mine a podcast or a vodcast?
2. What space am I going to use for my studio?
3. Is my studio plan within my budget?
4. Am I going to get all the equipment I need in the beginning or am I going to spread buying my equipment out over time?

4

Show Prep, Episode Mapping, and Script Writing (Noah Mears)

Now that we have some of the heavy lifting out of the way, let's begin to put our episodes into action. We'll start with what I like to call episode mapping. That is essentially creating a list of each episode for your next season. OK, I know this needs a little unpacking, just stay with me. When we talk about creating a list of each episode, this involves digging much deeper into your topic. Need an example? Great. What's that? You want an example of a season list on a podcast designed around BBQ? Even better!

Let's admit that BBQ is a very broad topic, probably similar to the topic you have chosen, so let's break it down. Understanding that I want my podcast to be informative and entertaining I'm going to choose to begin my first season with an informative episode. Also, I like to add what I call "slice of life" episodes. These are the feelgood, make the viewer reflect, possibly laugh or cry, just a little. Here's what I'm working on, we'll get a little deeper in each of these in just a second.

- Episode 1 The science behind that BBQ taste
- Episode 2 All your cuts of beef, chicken, and pork
- Episode 3 The secret really is in the sauce
- Episode 4 You're doing it wrong
- Episode 5 Smoke isn't smoke
- Episode 6 Life is a combination of magic and BBQ
- Episode 7 Give that man a rib!

- Episode 8 Good BBQ comes from experience, and experience, that comes from poor BBQ
- Episode 9 You gonna finish that?
- Episode 10 Circuit overload

Let's break down these episodes and why I've chosen them. The first episode is pretty straight forward, to anyone who has enjoyed good BBQ, you know the taste. This episode will answer just what that taste is. In so many words, we'll go into the science of the molecular structure of fire and how that impacts the flavor and the fibers of whatever it is you are cooking. This episode will be a solo hosted, informative, approximately 15–18 minutes long. Also, this episode will lay the foundation to information that will be delivered in some of the following episodes.

VasiliyBudarin/Shutterstock.com

IMAGE 4-1 Life is just better with good BBQ!

The second episode is another informative show. For the advanced BBQ enthusiast this episode may offer very little. However, for the persons who are just getting into the art of BBQ, this will inform those viewers of each of the cuts of beef, chicken, and pork so as to know what you are working with and how to treat each cut to maximize results. A local butcher will be the guest for this show. And for those of you who are keeping score, that's one show solo, and one show with a guest. The point of keeping count is to try and shake things up each episode. I wish to avoid keeping the same format with episodes, and give viewers another reason to check back in. This episode is probably going to be in the 20–22-minute range due to how much material will need to be covered.

The third show will change things up again. This show is all about BBQ sauce, and my intention is to line up three to four BBQ sauce chefs from around the country to discuss philosophies on what makes a good sauce, and how regional expectations have shaped the landscape in the development of popular sauces. This will be a panel discussion format, and the guests will connect via Zoom. A very light hearted, informative, entertaining episode is the objective here, and this should be in that range of 15–20 minutes.

For the episode "You're doing it wrong" I'll return to a solo show, discussing why your long cook items such as brisket and pork shoulder are coming out wrong. We'll go deep inside just what it takes to break down connective tissues and fats that result in tender, juicy BBQ. This particular show is going to require lots of graphics that I'll have to create before editing, so show prep for this one is going to take longer than most. This episode should fall within the sweet spot of 15–18 minutes.

Episode five is the halfway point of the season, and we'll really get into the dos and don'ts of applying smoke to your cook. We'll cover almost all of the woods available for smoking meats, we'll compare undertones and aromas each type of wood produces and how that couples with cut of meat. There's a lot to cover in this episode and a fellow BBQ enthusiast will join me for this 20-minute discussion on how smoke is anything but smoke, and they're all created differently.

It's at this point that I like to take a closer look at the season up to here. We are at the halfway point, what is working? What is not? Is viewership going up, down, or staying the same? How about social media reach, is it effective? Am I being consistent with posting? What about feedback from my viewers? Is it positive, negative, or nonexistent? Lots of questions need to be asked here, and if needed, trying something a little different for the second half of the season should be considered. Don't worry, Chapter 7 will address promotion and distribution, but for now just know that you must work equally as hard in this arena as you do in recording and editing.

Nick Starichenko/Shutterstock.com

IMAGE 4-2 We'll discuss all things ribs in the seventh episode

For the sixth episode of the season, once again, I like to change things up. This is the slice of life episode, the one of the season that isn't really informative, but personable. My "guests" for this episode will be a series of prerecorded interviews with my brothers, parents, children, even distant cousins where we talk about the good times in life where we gathered, laughed, cried, gave acknowledgement to lost loved ones, and did it all over great BBQ. The purpose is to have an emotional connection with viewers, and give an episode where, in some way, everyone can relate. You know, a slice of life. This one will be somewhat short, probably in the 15–17-minute range.

For the seventh episode, we'll jump right back into the flow of the show, and have a local BBQ rib master on as the guest. He's won professional competition rib challenges all over the country and there may not be a better qualified guest on such a topic anywhere in the world, and as luck would have it, he lives in the same town I do. This guy is a talker, and had a lot to say. We could go for just over 20 minutes with this one, and I believe no one will mind.

Let's take a quick tally of show formats. So far the plan is to have two episodes where I'm going solo, three with individual guests, one with a panel discussion, and one with multi-prerecorded sound. So far, so good. The format is designed to avoid getting stale, and most importantly, keep viewers coming back for more.

The eighth and tenth episodes will result from a series of single interviews designed to create two shows. I'll be interviewing professional BBQ competitors from a variety of circuits (California Circuit, Kansas City BBQ Society, International BBQ Cookers Association, etc.). The interviews will have guests explain some of their worst experiences while at competitions, and some of their worst results. This will be the portion that makes up episode eight. To maximize the use of the interviews, I'll also be asking about life on the circuit. The long days of driving and setting up. The thrills of getting the perfect results. The constant struggle to overcome the elements of weather, the cost of competition. The idea is to get deep inside life on the circuit overload. Getting two episodes out of one series of interviews is a good technique to reduce your workload and maximize content. Each of these episodes will attempt to get in right at the 17–20-minute mark.

The ninth and strangely enough, the final episode of this season will be my last solo, informative episode of season 1. Titled "Are you going to finish that?" I'll be sharing my tips, tricks, and natural recipes for viewers to properly clean, oil, and condition their smokers for maximizing the life and use of these units. While somewhat tedious, there is a method to gaining a natural season to your smoker, while avoiding leaks, breaks, and rust. This episode will probably be the shortest of the season at about 12–15 minutes, and will be full of graphs and still photography shots that will have to be gathered and created before production. Lots of time will be required for this portion, and gathering these elements should begin soon.

A_B_C/Shutterstock.com

IMAGE 4-3 A map can help almost every journey, your podcast season is no exception

So that's season one. Three solos, three individual guests, three panel discussions, and one slice of life. While this is by no means the written rule to how a podcast season should be planned, this is how I'm going to approach my first season with this new show. I encourage you to explore and move pieces around until you find the right fit for your first season. Don't feel like you have to be a master planner at the beginning. Get you a starting point, and get after it. You'll find what works along the way and what doesn't. Remember way back when we discussed perseverance? Now's a good time to revisit that, and get you a big ole slice of perseverance because we're heading into script writing, and then it's time to record!

Script Writing

Unlike a broadcast news production, where script writing for talent is essential, script writing for podcasts is very different. In fact, I personally only script the intro, otherwise I try to have the conversation very organic, and natural. I have a student at Tennessee Tech University, Noah Mears, who excels at many things, one of which is podcast production. He has his own unique way of approaching script writing for podcasts. Here, give it a read:

> Before you can hit record for your podcast, you must first decide what your episode is going to be about. This is why script writing and show prep are crucial to the flow of your podcast. One of the easiest ways to prep your host and guest with content for the episode is Script Writing. When I say this I don't mean write a 25-page script with intricate dialogue and slug lines (No need for slugs here). I am talking about building a simple checklist for your host to use as a way of navigating through the episode. Look at it like this, You are grocery shopping and forgot to make a list. Now, you know what you would like to eat but let's say you are cooking for a group of friends, and you want to ensure you have all the ingredients you need and stay away from the ones they might not like. The best thing you can do is ask around what dish to prepare and build a list. Once you have this list, you can now go shopping with confidence. This analogy applies to scriptwriting. You want to ensure that your host has a list of questions that flow in rhythm with your goals for your episode. While podcasting is a much more relaxed medium of storytelling, building a script is helpful and adds more cohesion to your episode.

Well said Noah, what we really need for a script is questions and talking points we want to make sure we get throughout the show. To expand on this, Noah Mears shares his four areas of focus that will define this list. They are:

Areas of Focus:

Background

- This covers their upbringing, childhood, college years, etc. Basically, you want to gather information about anything they have done prior to their current projects.

Current Projects

- This area covers anything they are doing currently whether it be work or hobby related.

Favorite topics

- This area relates to what brings them the most joy within their jobs and hobbies.

Future Plans

- This is anything they plan to do in the future! This content will be the ending portion of your episode. These four areas are going to support your entire podcast. Grabbing as many details about these four categories as possible will help you when fleshing out your script.

Again, very well said Noah, these are the four areas of focus that will make your questions in line with your guests. So how about some real examples of questions that fall into these areas of focus? I'm glad you asked. Here are actual questions used during a technology-centered podcast. The guest was a producer of virtual reality (VR).

Background

- How did you get started with VR productions?
- What drew you to VR?
- What was your first project?

VR Projects Overview

- How does a VR project get started?
- What goes into preproduction?
- Organization and management
- What are the different stages (alpha, beta, and client)?
- Postdevelopment
- Launch
- Updates and maintenance

VR's development

- How was VR at the beginning?
- Where it is now?
- What would you expect in the future?
- What are you most excited about in the future?

Current

- What are you working on currently?
- Any big plans in the pipeline?
- Where do you want VR to go?
- Favorite project?

These are a good foundation for your show. Are you going to ask each of these questions? Maybe, it depends on the flow of the show. Will you ask questions that aren't listed earlier? Probably, guests have a way of going off script and the natural discussion will most likely provide you an opportunity to ask a bunch of different questions.

Scripting the Intro

OK, we're getting closer. You'll be recording your first show in no time now. One more thing we need to do. Earlier I mentioned I like to script out the intro to each show. Well, it's important to know what that looks like and information included in each intro. Here is an example of an intro I am writing for an upcoming *Let's Go Smoke.*

> Welcome everyone to the premier episode of *Let's Go Smoke* where we talk all things BBQ … Got a brisket? … we can smoke that … Chickens? … We can smoke those too … If you can kill it, we can grill it … I'm your host Teddy Burch and today we're discussing everything you're tasting with your BBQ … And how you can get more of that taste you're craving … You love BBQ? You're at the right place for "Let's Go Smoke"

A few things about that intro, first it's relatively short, just under 20 seconds. The next elements it contains, let's get color coordinated for that one. Here it is again,

> Welcome everyone to the premier episode of "Let's go Smoke" where we talk all things BBQ … Got a brisket? … we can smoke that … Chickens? … We can smoke those too … If you can kill it, we can grill it … I'm your host Teddy Burch and today we're discussing everything you're tasting with your BBQ … And how you can get more of that taste you're craving … You love BBQ?, you're at the right place for "Let's Go Smoke"

The purpose of this line is to let your viewers know immediately what show they are tuning into. This is to let viewers know what they can expect from your podcast. This is an extended tagline, which is text used to clarify a thought. This line is to identify the show's host or hosts. This line is to let the viewers know what to expect on this episode. This line is the intro wrap-up and will be used for every episode.

These are the elements and the questions you are going to want to address with your intro. Be aware of the length of time of your intro. You should have intro music that fits the feel of your podcast. Usually, and you'll have to play with your intro to get the right combination for you, the intro format is as follows:

1. Intro music fades in, plays for 5–7 seconds
2. Music fades down, your recorded into plays
3. Music volume increases after you have completed your intro
4. Music only plays for a few more seconds, and then fades to mute
5. This will be the beginning of your episode

Again, you will have to experiment with your intro, and it needs to fit the overall feel of your podcast. It's a good idea to have your season 1 intro rehearsed before recording, and it's equally important to keep a consistent intro throughout your first season. If you feel it needs to be altered, avoid the urge to alter in the middle of the season. Work on it after you've completed season 1. Alright, it's finally here, it's time to start recording! See you in Chapter 5.

Chapter 4 Terms

Episode mapping Creating a list of shows for the entire season.

Scripting for episodes Creating areas of focus as pointers to use throughout the recording of your episode.

Scripting for intro Brief introduction to each episode with the purpose of letting your viewers know immediately what show they are tuning into, let viewers know what they can expect from your podcast, give an extended tagline, identify the show's host or hosts, let the viewers know what to expect on this episode, and to provide a wrap-up.

Chapter 4 Checklist

- I've read Chapter 4 and have a firm understanding of its context.
- I have a good understanding of how to map out my first season.
- I know who my guests are going to be for each episode.
- I have begun my script writing for my first episode.
- I have an intro that I like.
- I know exactly how long my intro is.
- I have music for my intro.

Chapter 4 Questions to Ask Yourself

1. Do I have each of my episodes mapped?
2. Do I have each of my episodes named?
3. Do I have my guests picked out for each episode?
4. Do I have a count of how my season is going to be formatted? For example, season 1 of Let's Go Smoke was three solos, three individual guests, three panel discussions, and one slice of life.

5

Interviewing and Recording Your Podcast

Well, here we are, it's recording time! OK, so we'll be recording in just a few. There are a few points to make about your recording. First, and you've probably picked up on this, and if you haven't I'll put it as clearly as possible. The shortest time span of anything related to your successful podcast is the actual time in the studio recording. Show prep and script writing, researching your guests, editing (Chapter 6), promotion and distribution (Chapter 7), and properly archiving will take longer than your recording time. To put it into another realm, I like to think of it as if you were on a football team. You will spend endless hours in practice, in the weightroom, running, etc. each week, and when the game comes on the weekend, it's 60 minutes long. If you're lucky you'll be playing for 30 of those minutes. More than likely it'll be less than that. Your prep time can be 10× that of your playing time. Your podcast is no different.

engagestock/Shutterstock.com

IMAGE 5-1 The time is here, it's time to record!

Plan now to spend lots more time on the other things associated with your podcast.

Next you got to get your guests lined up. I know, I know, prepare to be let down. Thankfully we've covered this in Chapter 1. We know it's going to happen, our guests sooner or later will be a no-show. We get it, we're prepared for it. Let's say this guest showed up on time and is ready to record. Greet them and walk them through the steps of what's going to happen during the recording.

Next, it is your job to make your guest comfortable. Since we are at this point in the podcast process, we've been working toward this point. Your guest, on the other hand, has not. By that I mean there is a good chance your guest has never been interviewed, or has ever been under the studio lights with the cameras rolling. If that is the case, again, it is your job to make them comfortable.

Why? Because a comfortable person will speak openly on your topic, an uncomfortable person will not. The best way to get your guests comfortable is to have a few minutes before the recording to "break the ice." Small talk over the weather, last night's basketball game, or even the topic you are about to discuss. The point is to make them relaxed and comfortable.

Finally, you have to be in the mindset that this is a one-take production. That is, when the record button is hit, there is no starting over. You'll have one shot to get the content you need. Sure there'll be editing time to get rid of the uhs and ums, the dead space, and the external noises that will inevitably be recorded. That's fine; however, the intro, the interview, and the outro is a one take production. Let's discuss these three elements of your podcast just a bit further.

Impactful note: It's important to know that the majority of this chapter focuses on principles for a vodcast (audio and video production). If you're strictly doing a podcast (audio only) the closest section will be the one host, one guest section. Obviously you will not need to follow my tips and tricks for a better on-air personality if you are audio only.

Recording Your Podcast

Earlier in Chapter 4 we discussed a few audio boards and switchers. Based on that information, and your research, you'll need to determine how you are going to record your podcast, and that is to say where the recording will be located.

As you begin to record multiple episodes, you are going to need a sizable amount of hard drive storage for rather large files. Make sure to check if your switcher has the ability to record (probably USB) to an external hard drive. If so, that is an option that will allow for saving and exporting to one central place, making it easier to keep track of all episodes.

If your switcher is unable or does not have this feature, you can use what is known as Network Device Interface (NDI) which is a royalty-free video over IP transmission protocol. NDI is designed to enable compatible devices to share audio and video over IP networks for broadcast production. NDI was developed by NewTek and to use their software you can simply go to www.newtek.com/ndi/applications and download the appropriate suite for your computer. Once downloaded, the Video Monitor application will be the program feed out of your switcher and will record audio and video as you are seeing it on your monitor.

Be prepared for some trouble shooting (Chapter 10), it has been my experience that anytime you are setting up these devices, there are always bugs to work out.

So now that we have guests, and we know where we are recording our podcasts to, let's discuss the actual shoe itself.

The Intro

Based on Chapter 4, you should have this written, and know what needs to be in your intro. The purpose for bringing this up here is to give you options. You can prerecord your intros before your guests arrive and edit them into your show later. By doing this option it gives you the freedom to have multiple takes in case of any mess ups. Or you can record with your guest and make the intro part of your one take recording. This is a fine option once you are confident, or have gained just a little bit more experience.

The Interview—One Host and One Guest

This is where the magic happens. This is the purpose of the entire process up to now. You have your guest in the studio, they are mic'd up, and the record button has been pushed. Excellent.

DC Studio/Shutterstock.com

IMAGE 5-2 It's all about the interview

As mentioned earlier, making your guest comfortable is key. Once they're comfortable, they talk. Next in line for the host to be aware of is pace. This is essentially how the story is being told to the viewers. The host must walk a line between moving the story along (not getting side tracked into off topic conversations) and making sure the viewers aren't becoming bored because the pace is too slow. So how does one keep pace? Excellent question! Let's jump into that now.

Interviews are made up of a series of questions. To truly get your guests to speak comfortably, ask open-ended questions. "How do you feel about ...?" or "What is your opinion on ...?" These are long-answer questions. There are several reasons for asking the open-ended question. They are:

- Get your guests talking. A talking guest is a comfortable guest
- The host and the viewers are getting lots of information. Hopefully good information.
- Their responses to your questions should lead you in the direction of additional questions based on their answers.

As the host, you have to develop a running clock in your mind that keeps a timer on your guests speaking. You have to develop a sense of when they are talking too long, or getting off topic, or even worse, repeating themselves. While your job is to host the show, part of that means directing your guests to tell their story in the timeframe your views demand. Experience is the very best practice for developing this clock.

So, how do you get guests that wander off subject back on track? Another good question, and the answer is simple. Close-ended questions: These are very direct and usually have a one- or two-word answer. "How old are you?" or "Have you eaten lunch yet?" These questions are designed to prevent your guests from continuing to speak. I know, I know that sounds like the opposite of what we are doing as the host and interviewer, however I think you're going to find that in general, people like to talk about themselves. Some people like to talk about themselves a lot. So, you have to keep them on track. Sometimes when you give a closed-ended question, and they are forced to give a very short answer, you have just been given back control of the interview, and this is the opportunity to get it back on track.

Stocksnapper/Shutterstock.com

IMAGE 5-3 You have to develop a timer in your mind to keep interviews on track

With that said, one of the main, most impactful practices you can adopt when conducting one-on-one interviews for your podcast is to convey to the guest they are the most important person on the set. Known as on-air intimacy, you want your guest to forget the cameras are recording, you want them to feel the bond and connection the two of you have over your very unique conversation. Choose your words wisely so they feel as though you are talking to them and only to them.

The Interview—One Host and Multiple Guests

The rules of open-ended and closed-ended questions are still the same for interviewing multiple guests, or even a panel interview. However, there are some unique elements introduced in this type of interview, and we need to discuss a few of those.

Sterling Munksgard/Shutterstock.com

IMAGE 5-4 Panel interviews are the same and they are different

The first and perhaps the most important element of interviewing a panel is to understand that every one of the members are different. Their backgrounds, their accomplishments, their research, etc., your first job in this setting is to do your research on every member before the interview starts! I cannot express this enough. You need to research every member until you are comfortable with a good amount of knowledge of each.

Again, you need to remember that all of the panel members are different. Which means you are going to get answers delivered to you differently. Some will give you replies in a happy tone, others will be the exact opposite, and some will give you replies with attitude. It's going to happen, just expect it. The main thing to remember is to not get off your interview objections, and stay focused on why you are interviewing this panel.

rangizzz/Shutterstock.com

IMAGE 5-5 No two persons on a panel interview are the same; be comfortable with different personalities

A couple more elements in the panel interview that are unique to this setting. The next is the on-air intimacy, don't spend a lot of time looking for this in the panel format, it just isn't there. In an attempt to find this rapport with one or two members, you'd likely break the next rule, and that is to be equal. That is to give each member equal on-air time, equal number of questions, and equal number of responses. You see, the moment you put your work out into the universe for critique, that's exactly what you're going to get. Everyone loves critique on somebody else. If you play favorites, or even slightly show favorites in a panel interview setting, don't worry someone is going to call you out and probably let you know how awful you're doing. Don't waste your time or energy in that cesspool (more on this in Chapter 7).

Try to remember what the great American journalist Barbara Walters said in regards to interviewing. "When you're interviewing someone, you're in control. When you're being interviewed, you think you're in control, but you're not."

The Solo Show

Remember when we talked about *Let's Go Smoke*, season 1, being made up of three single guests, three solo shows, three panel discussions, and one slice of life? Here we are going to address the solo shows.

diignat/Shutterstock.com

IMAGE 5-6 Solo podcast is perhaps the easiest of all the formats because you know all the information

When preparing for a solo show, remember you have to keep the conversion going on your own, so know your topic! I feel these are some of the easiest as far as researching for recording because you are (hopefully) the expert on what you are about to talk about. As long as you know your material, and you're confident in the pace you deliver your knowledge, you should be fine. Additionally, what I find is that the audience doesn't want to see my face for long periods of time so the most time in preparation for solo shows is making graphics, showing what I am saying.

Some Tips on Becoming a Better Interviewer

Like everything else, the more you interview, the better you get. Some of you have been interviewing for a while and have got the hang of it. Others have perhaps never interviewed anyone, for any reason. Either way, let's look at some tips that'll make each of us a little better in our next interview.

1. We've talked extensively about doing your research on the interviewee, this is for your preparation. However, once in the studio and the lights are on, and before your recording, prep your guests. Allow for time to provide them with some examples of questions you are going to ask. Let them get comfortable talking to you and sitting under the lights.
2. Have a goal in mind of where you want the interview to go, however start slow. Let your guests slowly work up to the more emotional answers. Remember this is an interview, you are looking for quality answers, have quality questions and don't go too fast with asking them.
3. Listen with care. Truly be in the conversation, and be aware that your body language is conveying a message to your guest. Make sure it's the message of truly listening to what they are saying and that you are interested.
4. Stay on point, don't get side tracked onto other topics. And if for some chance you do, have in your notes a reindeer to get back on track.
5. Nonverbal communication, learn the clues and pick up the messages. There's a good chance you're probably used to doing this. Keep working on it, for sure it'll come up during your interviews.

Olena Yakobchuk/Shutterstock.com

IMAGE 5-7 It is essential you learn the meaning of and pick up on nonverbal clues

A Few Tips on Having a Better On-Air Appearance

While some of these are obvious, and others may come off as not a big deal, here are a few tips for a better on-air appearance

1. Avoid checkered or striped clothing. There's a good chance these clothes patterns are going to clash with the set, and therefore clash with the viewers. Avoid creating any unnecessary distractions.
2. Sit up straight, there is no alternative. If you are slouched over, it's going to show. Not only is it going to show, it's going to be heard. Sitting up straight will allow for better airflow for you and will naturally provide you with the ability to speak clearer.
3. Avoid jewelry, it's going to reflect studio lights and naturally become a distraction for your viewers.
4. If it's a problem, a studio powder will reduce shine. If you are consistently working to edit or reduce facial shine, consider this option.
5. Turn off your phone, and have your guests do the same. It'll really kill the one-take philosophy. At the very least it'll create more time in the edit lab.
6. Finally, and let's hope this is never a problem, but never under any circumstance chew gum. That tip is hopefully self-exploratory.

Alright, so here we have arrived at the end of recording your show. As always, terms, questions, and your checklist are about to follow. Stick around for Chapter 6 where we'll discuss editing and getting the maximum quality out of your podcast.

An Audio Recording Checklist

- Prepare your notes. This is a must. Have you ever heard the saying there are no stupid questions? If you fail to prepare your notes and have a general direction to where you are going before you hit the record button, you are probably about to ask a lot of stupid questions.
- Check your equipment. Take a close look at your microphones, headphones, chairs, and desks. Are they firmly connected? Are they untangled? Is the area around where you and your guest are going to sit clean and not a distraction?

- Do a mic check. Make sure you are getting a clean sound from your mics. Also make sure your mic stands are in the correct location. Do this with headphones on to get an accurate test.
- Make sure your device that you are recording to is properly inserted and is working. More than once I've had guest on my podcast, probably had great conversations, only to find the memory card picked up nothing.
- If time permits, do a test recording. A test recording will expose potential problems such as hissing or a hum in the audio, or too much echo, or even natural room noise. Avoid learning about these problems in the finished product with a several second test recording in the beginning.

IMAGE 5-8 Using these checklists will help in avoiding problems in your final recording

An Audio/Video Recording Checklist

- Prepare your notes. This is a must. Stop me if you've heard this before. Have you ever heard the saying there are no stupid questions? If you fail to prepare your notes and have a general direction to where you are going before you hit the record button, you are probably about to ask a lot of stupid questions.
- Proper lighting. I always begin with turning on studio lighting. The purpose is to make sure the proper lighting temperature is present and that a nice, even throw of light is also present. This is also a good time to find unwanted shadows.

- Turn audio boards, switchers, computers and any other equipment on. At this point you'll need to decide if you are going to add logos, graphics and lower thirds during recording or in post-production. If using them live, load your graphics during this step.
- Set lighting, floor lighting, and monitors are turned on and the proper color lighting (if it applies) is selected.
- Turn cameras on and make sure lens covers are removed and camera is in the proper recording mode. This is the proper time to white balance the camera.
- Turn on any monitors that are on set and load any graphics you'll be using on those monitors.

  [Once talent and guests are on set.]
- Check your equipment. Take a close look at your microphones, headphones, chairs, and desks. Are they firmly connected? Are they untangled? Is the area around where you and your guest are going to sit clean and not a distraction or an eyesore for the screen.
- Complete a mic check. Make sure there is clean audio coming from the mics. The technical director will do this with headphones to pick up any wanted or unwanted noise.
- Frame your shots. Make sure each of your camera's subjects are properly placed within your shot. Think of your rule of thirds, or think of your shot as a tic-tac-toe board. Make sure on the close up shots of talent and guests and the top line of your tic-tac-toe board is running horizontally as close to the eyeline of your subject as possible.
- Manually focus the cameras. You might feel tempted to leave the camera on auto focus, resist this. When your subject moves, the camera is going to lose what it's focused on, if only for a second. As it tries to refocus, your views will see blur and most viewers have a very small tolerance for blur in their videos.
- Do a test run. Record just a little bit of audio and video, and play this back with the purpose of listening to any unwanted audio, shadows, poor lighting, poor video quality, slow frame rate, etc.

Poppy Pix/Shutterstock.com

IMAGE 5-9 Follow these checklists to ensure good, quality recordings

Chapter 5 Terms

Audio/video checklist A list of actions that should be taken before each recording to assure you are getting the highest recording quality possible.

Interviewing Asking a series of questions to a guest in order to get opinions or experiences.

Intro The beginning of your podcast that gives your viewer or listener an introduction to what they can expect from that episode of your podcast.

Nonverbal cues Transmission of messages or signals through the absence of speaking such as eye contact, facial expressions, hand gestures, posture, or the use of objects and body language.

On-air presence The fact or the condition of being aware of the image you are creating of yourself while in front of the camera.

Prep time All the combined time spent preparing for the recording of a podcast.

Solo Show Podcast format where you are the host and there isn't a guest.

Chapter 5 Checklist

- I've read Chapter 5 and have a firm understanding of its context.
- I have a good understanding of how to record my podcast.
- I know that I am either going to (a) prerecord my intros or (b) record the intros with the guest present.

Chapter 5 Questions to Ask Yourself

1. Who are my guests going to be?
2. Do I have necessary notes and questions for my guests?
3. Do I feel confident in operating the following?
    a. Lights
    b. Audio/video switcher boards
    c. Cameras
4. What's my comfort level at being an interviewer?

6

Audio and Video Editing (With Geoffrey Burch)

"The bottom line is that your performance is made in the editing room."
—Brent Sexton

Truer words may have never been spoken. It is vital that we understand it would be impossible to write *everything* about quality audio and video editing. However, this chapter is going to give you the basics for beginning, a few tips and tricks for saving you time, and with any luck the confidence to jump into editing and make the very most of your podcast.

Audio Editing—Let's Start Here

For this portion of the chapter, I'd like to give the floor over to a true expert. A person with decades of audio editing experience, and someone whom I've known for many years, my brother Geoffrey Burch. Geoff, the floor is yours.

Audio editing is a broad term that means different things to different people across a wide variety of industries and media, including television, radio, movies, music, and Internet broadcasts. This portion of the chapter will primarily discuss audio editing for the purpose of Internet podcasting. While many techniques and tools are similar, the function of audio from one medium to the next varies.

Audio editing is the process of manipulating recorded sound, whether it be an analog or digital medium, for the purpose of greater functionality, increased efficiency, and aural aesthetics. This portion of the chapter will discuss tools that affect volume, timing, and refinement of recorded sound, common issues and errors that arise within the process, and the methods to prevent or eliminate problems and increase workflow efficiency.

It is not uncommon for many producers to eschew audio editing as it is daunting and confusing. The task of this chapter is to explain audio editing fundamentally so that meaningful and productive work can be achieved effectively and efficiently.

The production stage is literally the recording of the audio. This is the process by which you will capture your audio and this can be done linearly, such as podcast, or nonlinearly such as ADR and foley for motion pictures. If you have set up your studio for proper audio recording, then the process of audio capture should go smoothly. With that said, there are some simple considerations to make while recording to further ensure a proper performance. While there are invariably issues that will arise in the recording that will require editing, a proper performance will eliminate much of the need to make any edits.

First and foremost, speak articulately. This suggestion is not specifically audio related, but how you and others speak will be the first thing perceived by your audience. There is no amount of audio editing that will repair uninspired and inarticulate dialogue. Literally every listener would prefer a mindful, well-thought-out speech as opposed to off-the-cuff slang and barely audible murmur. This might seem obvious but too often very well written content goes unnoticed because of the lackluster delivery of said content. The process of proper performance can be further augmented by being aware of your surroundings. Remember, your audience is not concerned with anything other than the relevant information you are delivering. Anything else is a distraction and can make the difference between multiple listens to multiple episodes or immediate turning off of the podcast.

While recording, avoid excessive breathing in proximity to the microphone. You may not even be aware that you are doing this, so be mindful of your exhalations. Also, get in the habit of not smacking your lips while talking. In day-to-day conversation this is not an issue, however every swallow and smack of your lips and mouth will not only be noticeable but amplified. Nobody wants to hear this so make sure you keep it to a minimum and have some water nearby in case your mouth becomes dry.

That brings me to the next issue. Make sure the objects around you are not causing extraneous noise. If you do happen to have water handy, great, but make sure you have a padded coaster nearby so that the microphone does not record the audible thump that occurs when setting a beverage on the desk. Anything that might make noise should be eliminated or handled with care. This will include pens, phones, tablets, and rustling paper. Specifically rustling paper is very difficult to remove from audio without degrading the audio signal. This is such an issue that on television and movies, paper items, such as bags, are actually replaced by a rubber prop to prevent the rustle of paper. The paper sound effect is added later.

Taking note of these issues during a well-rehearsed and executed podcast will result in a clean and smooth performance and hopefully eliminate any and all audio issues. Unfortunately, there are circumstances that do not allow for such mindfulness. Also, there are issues that arise that are not perceptible until after the recording has occurred, which takes us to the next phase of production.

Postproduction

During this phase you will be tasked with making edits to prepare your audio for proper broadcast quality. The hallmark of unprofessional audio production is inconsistent volume levels, obvious cuts and edits, and unwanted audio being present. I will go over the common issues that will arise even with the most well-planned audio performances. Firstly you will want to remove any and all unwanted audio. Most unwanted audio can be corrected by a clever use of equalization or EQ. Equalization is the process of adjusting volumes of varying frequencies within an audio signal. The first use of EQ that can be used immediately is a high-pass filter. A high-pass filter is an EQ curve that is used to reduce or remove frequencies below a specific cutoff frequency in an audio signal.

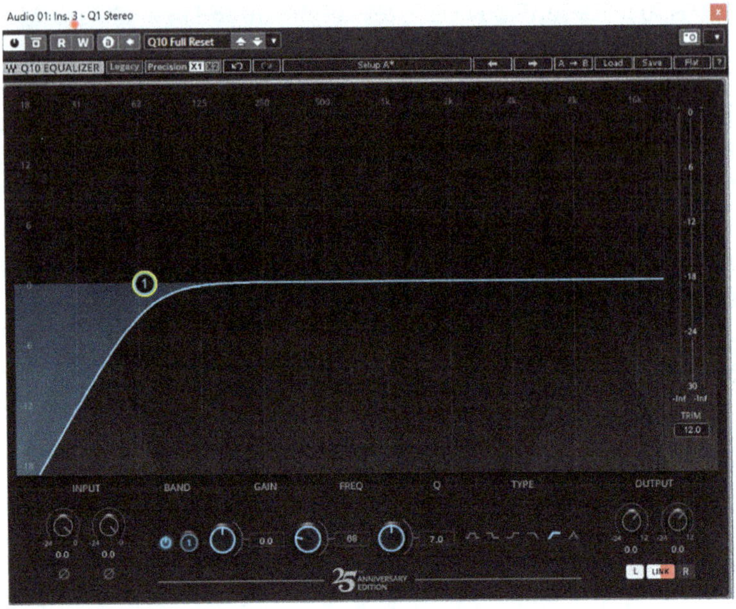

IMAGE 6-1 A high-pass filter

Human beings cannot typically hear below 20 Hz and so there is no reason to have any frequencies below that to be present in your audio. It is inefficient for speakers and will reduce the overall volume of your signal. For spoken word it is a good idea to set your high pass filter frequency cutoff to somewhere between 30 and 50 Hz. A high pass filter, properly set, is suggestible for literally all of your audio.

Plosives are when the air flow from the mouth is momentarily blocked by the lips closing, and then followed by a rush of air. This is most noticeable with syllables with the letter P or B. Even a well experienced podcaster will invariably run into this issue. Plosives are distracting and tend to be several decibels louder than the rest of the audio, limiting the dynamic balance of your signal. Plosives can typically be seen in a digital editor as obvious louder spikes in the audio.

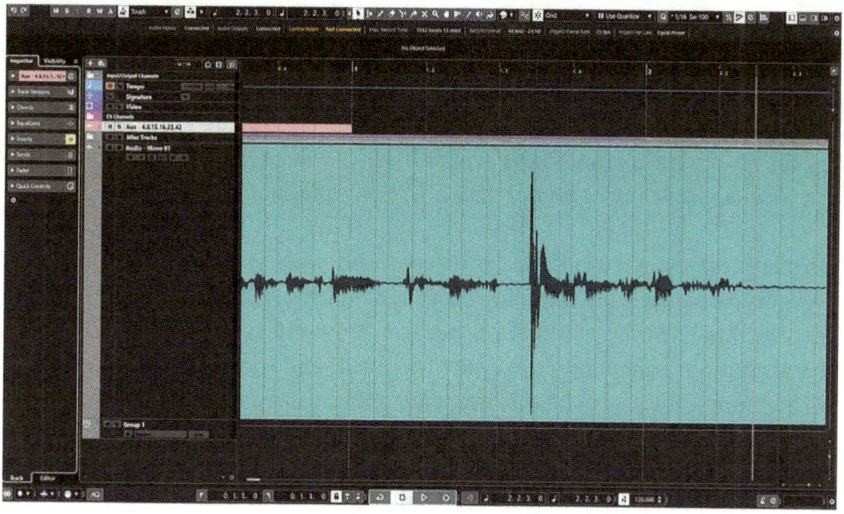

IMAGE 6-2 Plosives as seen in an audio track

Fortunately it is not terribly complicated to correct this issue. There are plugins available that specifically deal with removing plosives however most audio and video editing software programs have the functionality to remove them. Once again, a high pass filter with a cutoff around 120 Hz will remove most of the audible problem. Be careful and listen closely so that your high pass filter does not degrade the natural low end frequency of the audio signal. If that happens, it is possible to use a parametric EQ to locate and reduce the plosive frequency. A parametric EQ has three functions, Gain, Q Factor, and frequency. The frequency function allows the user to adjust which frequency will be affected. The gain will allow the user to increase or decrease the volume of that frequency. The Q Factor will allow the frequency adjustment to be more narrowly precise or broad, spreading across a wider frequency spectrum. Most EQ plugins will allow for spectrum analysis of the audio, which will allow you to see the offending frequency. Even with that function, it is prudent to rely on your ears and engage in some trial and error.

Another issue that occurs is that of sibilances. Sibilances are high-frequency hisses that occur when using syllables such as S and sometimes T and Z. For this we are going to look at what is called a DeEsser, which is widely available with most editors. A DeEsser does exactly what it sounds like; it reduces the frequencies where the "S" Syllable lies. Sibilances tend to sit in the 5000 Hz range which means that most default settings will solve the issue. If they do not, the DeEsser will typically have controls that will allow you to locate the offending frequencies and reduce it to a proper level. The "S" syllable is a naturally occurring sound so it is not advisable to completely remove it. Once again, use your ears to find the proper balance. Much like Plosives, a parametric EQ will work if you do not have a DeEsser plugin.

While plosives and sibilants can be reduced with proper performance and plugins, some unwanted audio comes for unforeseeable reasons that are not perceptible until the editing process. I recommend editing your audio using both studio speaker monitors switching back and forth with a proper set of headphones. If you can hear issues from either source, your listeners will also.

One common issue is a consistent buzz. A buzz in your audio signal is usually the result of electrical problems such as grounding issues, electromagnetic interference, improper shielding, poor wiring, ground loops, and proximity to strong magnetic fields. These issues are usually avoidable but they can creep into even the most robust studios if not careful. Because this is such a common issue most audio and video editors will have plugins to eliminate this. They are often titled as DeHum or DeBuzz or DeNoise, and they tend to work quite well. Unfortunately there are circumstances that do not allow for a complete removal of these frequencies, so as much as possible, troubleshoot this issue. Very often buzzes will often lie in the 120 Hz area. The process of removing these frequencies also lies on an EQ. High-pass filters will go a long way in reducing this issue.

While buzzes are often avoidable, one issue that is less avoidable is a hum. While a hum can result from an electrical issue, they are more often introduced by a source external from the audio source, usually from AC ventilation. Hums are typically a low frequency at 60 Hz. Once again, a DeHum plugin will work as will a high-pass filter or notch filter.

Once you have removed hums, buzzes, pops, smacks, and sizzles, you will almost always need to edit the timing of the audio. This will include removing unnecessary empty spaces. While listening to podcast audio, unnecessary dead space will dull the experience and it is necessary to remove this space. Many podcasts are constrained to a specific length and the necessity to remove time is inevitable. The issue that comes from audio and video editing is the audible cut. When an audio signal is cut, the audio signal is immediately set to 0 resulting in an obvious cut in the audio which is very easily noticeable and jarring for the listener. In this case, it is necessary to do an audio crossfade between signals. This is the process of one audio signal fading out volume as the next audio signal is fading in. With a proper crossfade these cuts will be made seamless. Audio and video editors allow this function and are easily executed.

Regardless of the room you are recording, the microphone is recording the tone of the room. Room tone can be affected by the shape of the room and the surfaces present. This is unavoidable but is also usually imperceptible in most circumstances. When cutting audio however, a change in room tone is perceptible. For this it is a good idea to record several seconds of audio of just the empty room. This room tone sample can be used to supplement the room tone that can be removed in an audio crossfade.

Once you have removed unwanted audio and properly edited the content, it will be necessary to boost and/or reduce audio levels proper for proper broadcast. Even with proper recording with proper levels void of extraneous noise, an audio signal may sound dull or lifeless. Some microphones and studios have a frequency response that leaves the recording sounding less than stellar. Once again, EQ can remove unwanted buildup of frequencies or enhance a lack of frequencies. The human voice resides within a limited spectral range so the removal of certain frequencies is optimal. It is not uncommon to cut low frequencies from around 100 Hz and below and to boost higher frequencies around 1000 Hz to brighten the overall timbre of the voice. Once again, use your ears to determine what you think sounds best.

From here, the audio recording is sounding good, void of problems. Even at this point it is necessary to make your audio consistent so that your listeners can consume your product without being distracted without varying audio levels. This is likely the most important issue of your entire audio signal.

To give you an idea of what I am talking about, think about a time you were watching a television show that was interrupted by an advertisement that was obviously louder than the program. This is an often talked about issue that is the result of inconsistent audio between the program and advertisement. Nobody likes this and with proper techniques you can avoid this altogether. The technique involved with this is compression.

Another important aspect of volume adjustment is compression. I want to take an aside and clear up the difference between audio compression and audio format compression. Audio compression is the technique of reducing the dynamic contrast within an audio waveform whereas audio format compression is the computational technique of reducing the file size of an audio file with minimizing a reduction in quality. An example of this is converting a .WAV or .AIFF file to a .MP3 file. While converting to a lower quality file format will potentially affect the way audio might sound, it has nothing to do with the very necessary technique of audio compression.

I mentioned that audio compression is the reduction of dynamic contrast within an audio waveform. What exactly does that mean? Take a look at Image 6-3. What you are looking at is a digital representation of an audio signal.

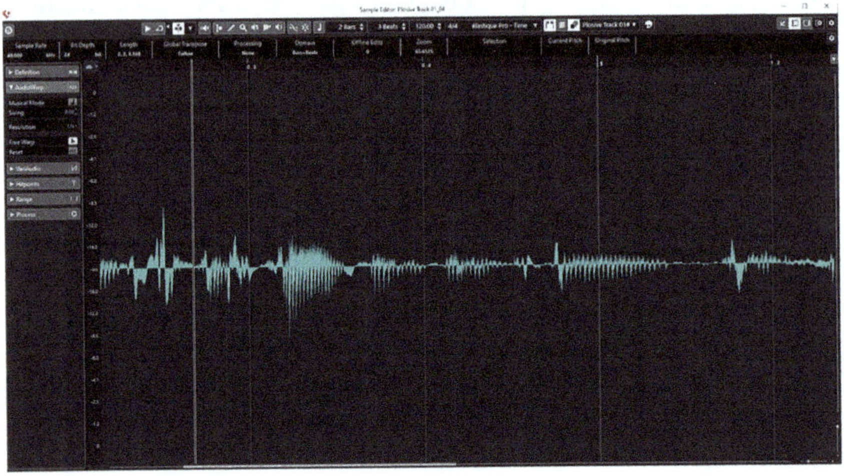

IMAGE 6-3 A basic audio signal

Looking at Image 6-4 is a zoomed-in version of that very audio wave form. You can see the frequent changes in amplitude. The peak, as marked by EXAMPLE A is the loud part of the audio and EXAMPLE B is the quiet part of the audio. Audio compression typically reduces the loud part closer to the quieter part, that is reducing the distance between the loud and soft moments, thus reducing the dynamic contrast. The purpose of this is so that you can boost the entirety of your audio so that it transmits in a way that does not necessitate the listener to adjust volume. That way, nothing is too soft nor too loud.

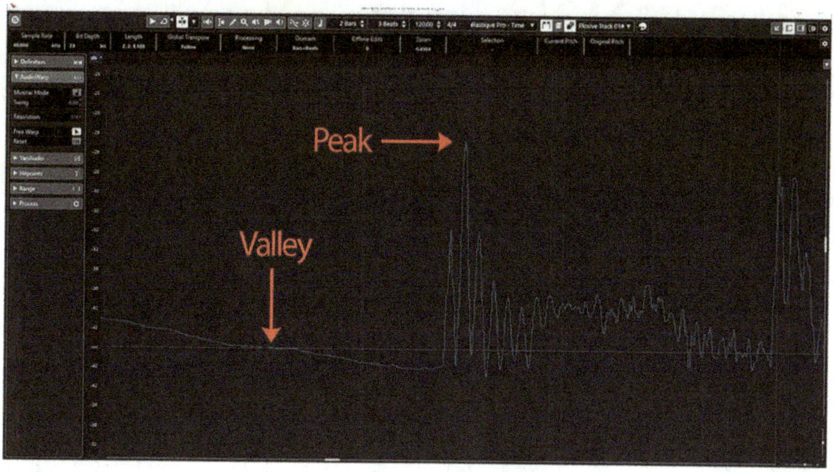

IMAGE 6-4 A zoomed-in view of the loud part of the audio

So how loud should the audio be boosted?

This answer to this question changes more than one might imagine. As digital technology grew the ethic was to make the audio loudness as high as possible. It is important to differentiate between loudness and volume. Volume is the scientific measurement of sound, usually measured in decibels or dB. A description of volume is quantifiable. Loudness is a qualitative term based upon the perception of the sound's intensity, usually measured in loudness units or LU.

Different platforms have different loudness unit specifications and oddly many of these platforms do not necessarily publish what those specifications are. If you are thinking about a specific platform, check to see if this information is available. If not, the generally agreed upon standard n is –14 LUFS (loudness units referenced to full scale) with a peak volume of –1 dB. If this is difficult to grasp and sounds like jargon, that is because it is.

Fortunately there are plugins and even websites that allow us to measure these numbers and master our audio to this standard. In most editors the plugin will have a name such as "metering" or "loudness meter." These plugins will give you information such as the integrated loudness meter and the peak value. It may take some trial and error, but adjusting the master channel compression and volume will get you to your desired levels. It is important to strive for these levels because the online services you use will automatically set your audio to their standards. The upside to this is that your audio should come through loud and clear on any given platform. The downside is that a platform's remastering of your already mastered audio could potentially diminish the dynamic range and quality of your work.

I hope this brief overview of audio editing gives you some tools to better your podcast. With that, I give the floor back to Teddy.

Video Editing

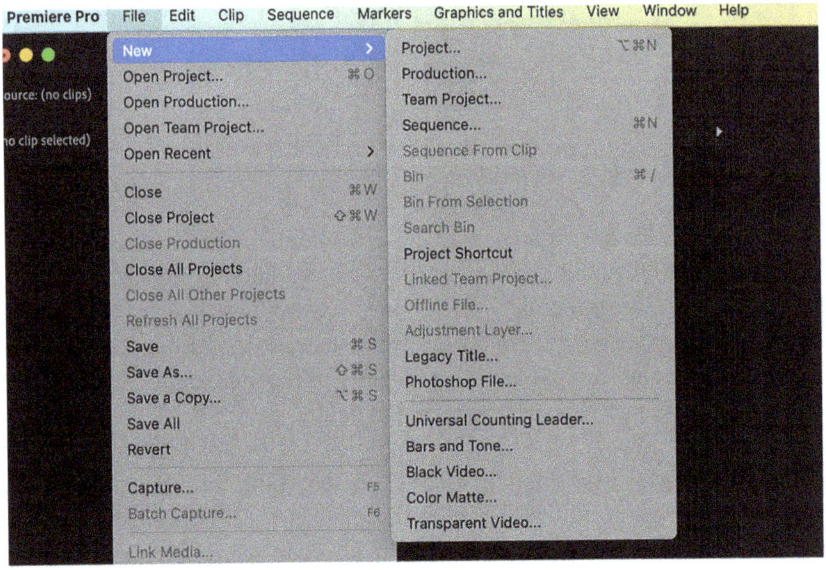

IMAGE 6-5 Opening a new project in your editing software

As we begin this portion of the editing process, I would like to reiterate that there are several different video editing software programs available. We are using Adobe Premiere Pro for the purposes of this book, however that is not to say it is superior to other editing platforms, it is simply the one I have used for many years and the one I am most comfortable with. Additionally, most editing software is similar in look and in tools for editing.

To begin, we simply want to open our recording as a new project.

Once open, you'll see your video in your timeline panel.

IMAGE 6-6 The preview video in the upper right-hand corner and the video in purple in the timeline panel

The duration of the video is indicated at the top of the timeline panel, and you can zoom in or out to adjust the duration intervals by adjusting the circular bars located at the bottom of the timeline panel. Similarly you can adjust the size of the video indicator by adjusting the circular bars to the right of the timeline panel.

Let's go over the tools that you are seeing and their function. First the lock next to the V1 or the first track of video is your toggle track lock. You can lock this track to prevent accidental changes to your video track.

To the right of the blue V1 indicator you have the toggle sync lock and the toggle track output. The toggle sync lock is enabled by default on all tracks and ensures that those tracks are affected equally by certain adjustments made to any track. The toggle track output allows you to isolate individual tracks to troubleshoot, or listen to individual tracks without the others. This is useful when you begin to have multiple video tracks.

IMAGE 6-7 The toggle track lock, the toggle sync lock and the toggle track output

To the left of the timeline panel, there is a set tools. Let's identify each one with their function.

IMAGE 6-8 The set of tools you'll use for editing your video

From the top, the first tool is the selection tool. It allows you to do basic selections, edits and movement. The second tool is the track select forward tool. This tool allows you to select a point on the timeline and move all clips that are before or after that point in either direction. The third tool is the ripple edit tool. This allows you to lengthen or shorten a clip while moving the rest of your timeline to accommodate. The fourth tool is the razor tool, and this allows you to cut or split clips to isolate the parts of video you wish to keep or delete. The fifth tool is the slip tool, and this allows you to change which part of a clip to display while keeping it the same size and length on the timeline. The sixth tool is the pen tool, and this allows you to draw straight, or curved paths. The seventh tool is the rectangle tool which allows you to draw a variety of shapes. The eighth tool is the hand tool, which allows you to grab and move clips, graphics, and other elements to your timeline. The final tool is the text tool which allows you to type text over your video.

Next, let's create some bins in order to maintain a clean, easy to find system for when we begin to add multiple audio and video tracks, along with graphics, music, etc.

IMAGE 6-9 The project panel where you will create your bins

Simply click on the folder at the bottom of the project panel and create a new bin. Give it a name, such as graphics, and place the appropriate files in the appropriate bin. Create as many bins as necessary to house all the files needed for your project.

IMAGE 6-10 Four bins to house all the files needed for a typical podcast.

The upper right corner is your preview panel. This allows you to see the progress of your project.

IMAGE 6-11 The preview panel

The upper left panel is your source monitor, it allows you to view individual clips from your timeline panel aside from the preview panel, which shows you the project with all clips. Simply drag and drop individual clips here to view. This panel also lists timecodes and allows for editing effects.

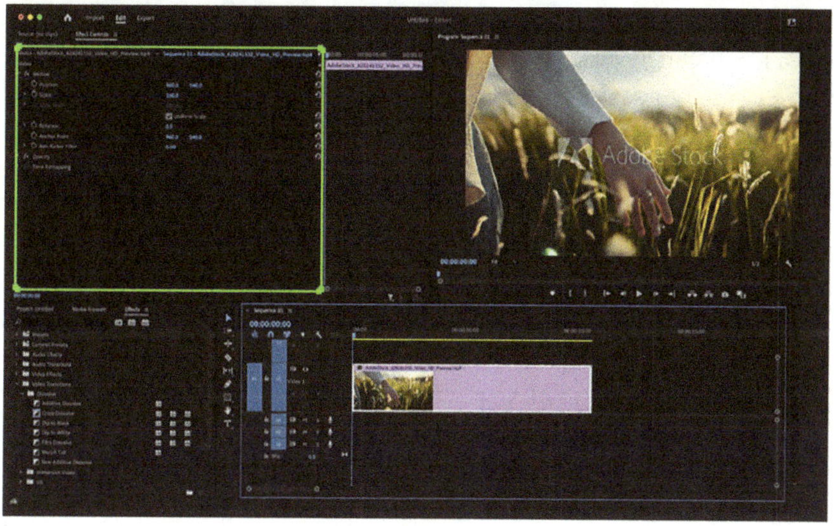

IMAGE 6-12 The source panel

Now that we have identified the panels and tools, along with their functions, let's edit some video and add some effects.

Trimming

Once you have video clips in your timeline, the space bar becomes the stop and pause button. View a portion of the video to make sure it is in timeline with your audio. One of the most used tools is the razor tool to make cuts and delete. Simply select the video clip you wish to clip, then using your razor tool, position it over the clip where you want to make a clip and click. The clip is now cut. Repeat this at the end of the portion you wish to cut out and then select that portion and hit delete.

To reposition a clip, select it and drag it until it touches other clips, removing any gaps. You can also use your ripple effect tool to trim the edge of a clip and automatically close the gap with the clip next to it.

You can also trim clips be selecting the clip, hover over the selected clip until the trim icon appears and then select and drag the edge of the clip to the desired length for clipping.

Moving Clips in Your Timeline Panel

You can drag a video clip and place it in an empty spot or snap it to another clip for smooth transitions. You can also insert and overwrite clips that you move. Watch the translucent rectangle that represents the clip's duration as you drag it. You can also move multiple clips by selecting more than one clip and moving in sequence to the desired location.

Adding Graphics

There may be a portion within your project you want your views to see full screen graphics (that is a graphic that takes up the entire screen). To do this, drag and drop the graphic over your clip in V1. Adjust the duration of the graphic by selecting the graphic and adjusting left and right to shorten or extend time. Play back the video and determine if the transition is smooth or does there need to be a video effect placed in order to make for a smoother transition.

Video Effects

There are many, many video effects you can add to your project. However, let me encourage you to think of video effects like cayenne pepper, a very little bit of it goes a long way. In the panel where your bins are located, the top will have three different tabs. These are Project Title, Media Browser, Effects. Click the Effects tab and you should see six folders titled Presets, Lumetri Presets, Audio Effects, Audio Transitions, Video Effects, Video Transitions.

IMAGE 6-13 "Effects" folders' location and names

Each of these folders will have a set of additional folders, and with those additional folders are the files to give your clips different effects. The one I use the most is located in the Video Transitions, Dissolve, Cross Dissolve.

IMAGE 6-14 The location of the video effect Cross Dissolve

Simply drag the cross dissolve file from the folder and place it in the desired location in your video clip. These are typically used in transitions at the beginning of clips, or the ending of clips, or from video to graphics and back to video.

To get more familiar with effects and the effect they will have on your clips, take the time to drag and drop multiple files, one at a time, over to your clip and watch the effect. You can undo the last move with the shortcut control + Z for PC and command Z for Mac. Keep a list of the ones you like and be creative in the ways you use them.

Exporting

Once you have edited your audio, video and any graphics you wish to add to your podcast, you'll need to export your project to share on a variety of social media platforms. To save and export is a simple process and most of the default settings are going to work for most channels you wish to share your podcast too.

To get started, at the very top of your editing software interface, click on file, then export. Here is the interface for Premiere Pro.

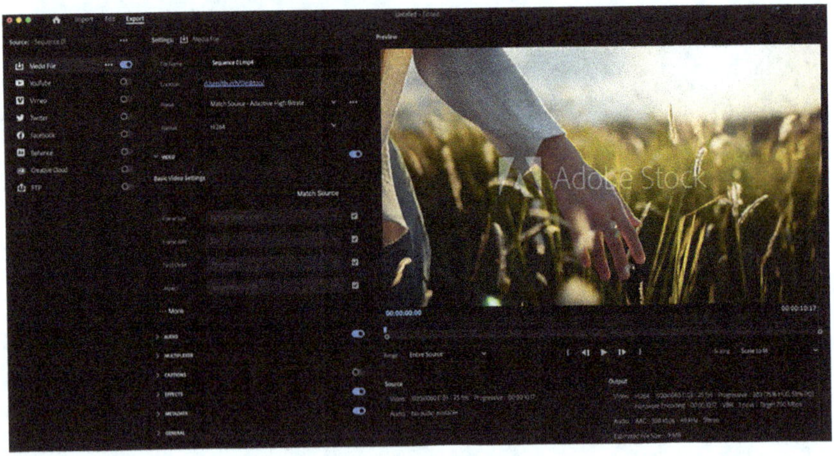

IMAGE 6-15 The options for exporting your project

The first panel within the exporting options allow you to share directly to a variety of social media channels. This option is not available on all editing software programs and each account will have to be linked to you Creative Suite account in order to use this feature.

The second panel allows you to choose the file name, where the file will be saved too. The third option in the second panel allows you to choose your preset. The default here is match source—adaptive high bit rate: I use this one almost every time. By selecting this option over HD or 4K allows for your video to adjust to whatever Internet connection your viewer has and your video will automatically adjust for seamless viewing.

The fourth option in the second panel is default H.264 and again I always use this, as it is a video compression standard known as MPEG-4 Part 10, Advanced Video Coding, MPEG-4 AVC, or AVC video.

The third panel supplies lots of helpful information. Project source, size, duration, frames per second, audio settings, and file size can be located on this panel.

When you are pleased with the results of this page, click the export button and your video will be compressed, and exported to your set location. In the H.264 preset, this mp4 is now ready to be shared on all of your social media channels.

And that, as they say, is that. A brief introduction into editing your audio and video for your podcast. As with anything the more you work with whichever software you choose, the better and more comfortable you'll get. As stated earlier, most of the software programs available for editing audio and video are similar in their interface, and are designed to accomplish the same objective. Be sure and stick around for Chapter 7. You're going to want to learn some ways to promote your newly edited podcast!

Chapter 6 Terms

Adaptive high bit rate Allows for your video to adjust to whatever Internet connection your viewer has and your video will automatically adjust for seamless viewing.

Audio compression The technique of reducing the dynamic contrast within an audio waveform whereas audio format compression is the computational technique of reducing the file size of an audio file with minimizing a reduction in quality.

Audio editing is the process of manipulating recorded sound, whether it be an analog or digital medium, for the purpose of greater functionality, increased efficiency, and aural aesthetics.

H.264 A video compression standard known as MPEG-4 Part 10, Advanced Video Coding, MPEG-4 AVC, or AVC video.

Hand tool Allows you to grab and move clips, graphics, and other elements to your timeline.

High-pass filter An EQ curve that is used to reduce or remove frequencies below a specific cutoff frequency in an audio signal.

Pen tool Allows you to draw straight, or curved paths.

Plosives when the air flow from the mouth is momentarily blocked by the lips closing, and then followed by a rush of air.

Program panel The portion of the editing software that shows you a preview of your project.

Project panel the portion of the editing software interface that contains bins, searches for files on your computer or server, has links to effects

Razor tool Allows you to cut or split clips to isolate the parts of video you wish to keep or delete.

Rectangle tool Allows you to draw a variety of shapes.

Ripple edit tool Allows you to lengthen or shorten a clip while moving the rest of your timeline to accommodate.

Selection tool Allows you to do basic selections, edits, and movement.

Sibilances High frequency hisses that occur when using syllables such as S and sometimes T and Z.

Slip tool Allows you to change which part of a clip to display while keeping it the same size and length on the timeline.

Source panel The portion of the editing software that allows you to view clips individually, and stores multiple clips in one place.

Text tool Allows you to type text over your video.

Timeline panel the portion of the editing software interface that stacks audio and video clips for arranging, and editing

Track select forward tool Allows you to select a point on the timeline and move all clips that are before or after that point in either direction.

Chapter 6 Checklist

- I've read Chapter 6 and have a firm understanding of its context.
- I have a good understanding of how to begin editing audio and video for my podcast.
- I know the editing software I am going to use.

Chapter 6 Questions to Ask Yourself

1. Have I applied some of the mentioned effects to an audio track within an editing software?
2. Have I applied some of the mentioned effects to a video track within an editing software?
3. Have I exported my edited file into the format that is best suited for distributing my podcast?

7

Social Media Promotion and Distribution

Well, you've made it this far, congratulations! No, really, you deserve it. If you look back at how far you've come, this is a good time to realize you've got what it takes to make this work, maybe. I say that because now is the hard part. I reiterate this to my students all the time, the promotion of your podcast is the single biggest time commitment, and perhaps the most important.

If you think about the way you consume social media, be honest, if you are anything like me, you are a consumer. That is, you scroll through your favorite platform for the purposes of being entertained. Nothing wrong with this, in fact I feel as though this was the purpose of social media apps to begin with.

fizkes/Shutterstock.com

IMAGE 7-1 Scrolling through social media

What we need to do now is change our mindset on our purpose in social media communities.

In just a minute we're going to identify some things you are going to want to avoid as you begin planning your strategy, but first on the topic of mindset, we need to begin thinking of ourselves as marketers with a purpose and a plan. We post and we share because each post has a specific purpose. We need to begin thinking of ourselves as members of the community.

Or, let's put it like this: "Engage, Enlighten, Encourage and especially...just be yourself! Social media is a community effort, everyone is an asset," as stated by Susan Cooper.

Cannot overstate how important engagement is to building your successful online presence. Viewers are going to want to know you are caring and passionate about your podcast. Show them that you are.

OK, so let's address some of the biggest mistakes that can be made when designing your strategy.

Five Myths That Simply Aren't True About Social Media Marketing

All Social Media Platforms Are the Same

Nope, not even close. As I'm sure you've probably realized, TikTok is not Facebook, and X is not Instagram. Their audiences and messages are completely different, and your postings need to be custom made for each. Remember, you goal is to get into the community and add content and discussion into that community. Engage with followers and they will engage with you.

You Need to Post on Every Platform Every Day

Again, not true. In fact this practice is a good way to runoff your potential audience. Rather, your approach could look something like posting every day to one or two platforms. This feels like a good point to clearly state, if any single person is claiming to fully understand the very best practices to successful social media marketing, they may be right today, but that's going to change really fast. This type of marketing is changing every day. That said, I've always liked the three-to-one posting ratio. When you are promoting your podcast, that post is easily recognizable. So, let's say your podcast is political based. Post one might share some recent news about a political candidate, not gossip, but real news. Your second post might share the overall results of a recent Gallup poll. The third might be information about an upcoming debate, followed by the fourth that is a promo for your podcast. These of course are not all posted in a single day, but over the course of several days.

Rawpixel.com/Shutterstock.com

IMAGE 7-2 The purpose of your social media strategy is to get people engaged in your content

You Don't Need a Defined Promotion Strategy

Couldn't be further from the truth. Without a clear plan, your promotion will fade away into social media abyss. If you choose to use the three to one approach (or whichever plan works for you) using your notes, or calendar in whichever email service you use, plan out your two or three week strategy. Be specific about what you are wanting to post on which days. Be flexible as things have a way of changing, but be consistent. Know why you are posting on a particular day, as well as what you are posting. And don't forget to add in "Slice of Life" postings. Followers like to know that you are a real human, with real human thoughts, feelings, emotions, etc.

Analytics Are for Someone Else

Nope, they are for you. There is a tremendous amount of data associated with your accounts on various platforms. You need to be taking advantage of this. As your audience grows you can gain valuable knowledge such as engagement rates, reach, views, impressions, and click through rate, which measures the amount of people who click on a link in one of your posts to access your additional content. Make time to analysis your data, and allow the data to guide you when planning your next block of posts.

Never Pay for Promotional Posts

I believed this once upon a time. I'm certainly not here to promote big tech and their advertising department; however, I am here to say you should consider creating a budget for paid promotions. They will increase your reach and make a bigger impact for your community. They will also be the outlier in your analytics, so use paid promotion like Five Spice, sparingly.

Five Tips for Better Social Media Promotion

Practice Social Listening

Listening, it truly is a dying art. But here's your opportunity to do just that, and maybe even increase your reach organically. Social listening is the act of observing what the discussion is within your community. That is something that is trending. Perhaps it is something political that you are unaware of. The idea is to hear see what your community is talking about, informing yourself on this topic, and posting toward this topic with educated and informed information. You may have a podcast on politics, however you cannot possibly know everything political. This allows you to become a better informed promotor.

Don't Forget the #Hashtag

I know it's no longer 2008, a time when hashtags ruled the social media world. And while they are no longer king of the hill, they still serve a purpose. Let me first acknowledge the opposite side of this argument that hashtags are simply not needed any more, and while they may not have the same reach they once did, they are still relevant. You can remain in the conversation with the use of them. Check out Facebook, YouTube, X, you'll see they're still there.

Khosro/Shutterstock.com

IMAGE 7-3 Hashtags are still a thing, use them

Take Advantage of Scheduling Posts

If you are anything like me, the majority of your weekdays are busy, and scheduled from morning to evening. This makes it easy to push back posting until the day is late and it becomes a forgotten task. Avoid this and schedule posts (where applicable) in advance. Most platforms have an option to schedule posts within the dashboard, if you cannot schedule in the application itself, use third-party software programs. Hootsuite, Sprout Social, and Buffer are just a few of the available scheduling software programs.

Take Full Advantage of a Detailed Profile

Believe it or not some social media users make quick decisions on whether or not they like you. Sometimes, all it takes is an incomplete profile and the decision is made. Most platforms allow for an extensive profile. Take the necessary time to add as much information about yourself, your podcast, and other related info to give to your viewers. Just remember, be yourself, everyone else is already taken.

Ignore Your Total Number of Followers

Alright, I get it. To completely ignore the total number of followers you have is impossible. Perhaps we should say spend very little time on your total number of followers. Instead of worrying about the total number of followers, focus on giving the number you do have the very best content you can create. If you are consistent, the audiences will come.

These are some tried and true ways that work in promoting yourself and your podcast via social media platforms. You may find you have to alter your posting from time to time. Remember, the social media marketing arena is always changing, be vigilant about changing with it.

Now that we've discussed promotion, let's talk about distribution and specifically where should I house my podcasts, and what can I expect during this process.

RNDSGN STUDIO/Shutterstock.com

IMAGE 7-4 Just in case you need the encouragement

Where to Store Your Podcast

Anchor.fm is a one-stop shop for all your audio only, or audio and video podcasts. The platform allows you to create an account, upload your files, share with just about every podcast station, monetize your podcasts, and provide analytics. It's a user-friendly dashboard that supports mp3, m4a, wav, mpg, mp4, or mov file types, and at the moment still has a robust free version that'll help get your podcast promotion off the ground.

* Remember this? That's right, it's where I put my opinion. Anchor.fm is a wonderful spot for holding your podcasts and for publishing them to a variety of podcasting sites. I would use Anchor.fm in one of two ways. If your podcast is audio only, you're golden, it's a terrific place for those files. If your podcast is both audio and video, I would take the video tracks off in the editing portion, and use only the audio files for uploading to Anchor. For uploading your videos, I'd still go to the reliable, and incredibly popular YouTube.

If you have a YouTube account, it's very simple to upload, share links, create thumbnails, get analytics, and store your video podcasts. At first it may feel like no one is seeing your work, and the site is over saturated with content similar to yours. Keep going, remember why you got into podcasting, and let that fuel you. I've been following certain channels for years and they had very few followers for the longest time, and then magically hundreds or thousands of followers appeared. No one is an overnight success on any of the social media platforms, you just have to keep after it every day.

Let's Bring It All Around

The first thing is to gauge where you are in the world of social media promotion. Perhaps you've lots of experience and are ready to jump right in and get to posting, if so, that's excellent! Perhaps that's not you and you occasionally scroll through your social media feed, but never with the intention of purposely gaining followers and views. No worries, we're going to create a plan now.

Rawpixel.com/Shutterstock.com

IMAGE 7-5 Social media demographics

First things first, create an account for your podcast, for each of your social media channels. Those are:

- Facebook
- X
- Instagram
- YouTube
- TikTok
- Anchor.fm
- Vimeo

The next thing I'd encourage is do not solely use your personal accounts for your podcasting promotion. Sure, it's fine to promote yourself on your personal accounts, but create unique accounts for your podcast on each of the platforms listed here.

Next, avoid falling into your own mental trap. Often, I hear students say I'm not on such and such app because I don't like it and don't want to bother, We have to lose that mentality, and put our product in every channel possible.

Know the audience on each platform. This is a changing dynamic and I would say it's a good idea to keep up with this regularly. As of the Spring 2024, here are some demographics we know. Facebook has just shy of three billion users worldwide, the largest age population is 25–34, and the site is made up of 56% male.

Instagram has roughly 1.2 billion users, the largest age population is 18–24, and is almost split evenly as Instagram users are 51% male and 49% female.

TikTok has self-reported 835 million users, the largest age population is 18–24, and the majority is female at 53%.

X has 240 million users, the largest age population is 18–29, and is 60% male.

Just knowing this much about these platforms should begin to give you a plan for what you are going to post on each platform. Just remember, Facebook and X are great platforms to be in the community and engage with members. Instagram, YouTube, TikTok are great places for reels or shorts, small snippets of your podcast to peak interest. Anchor.fm is a great place to upload audio files and share with multiple podcasting sites, and Vimeo is the alternative to YouTube and preferred by many.

ronstik/Shutterstock.com

IMAGE 7-6 Social media strategy

Now let's know the tips from this chapter, create all our accounts, and start strategizing your social media promotion plan! It's good to plan a short-term plan with obtainable goals. Be real with your goals, and sooner than later you've got to start posting.

Chapter 7 Terms

Analytics Information resulting from the systematic analysis of data or statistics.

Anchor.fm Anchor.fm is a one-stop shop for all your audio only, or audio and video podcasts.

Community Members of a social media platform that share common interests.

Engagement Term for actions that measure to what extent your audience interacts with your posts.

Promotional posts Social media posts that you set a budget for, and they are designed to reach a far greater audience than an organic post.

Social listening The act of observing what the discussion is within your community.

Three-to-one posting ratio The practice of posting three posts related to your podcast, or the topic of conversation within your community to every one post promoting your podcast.

Chapter 7 Checklist

- I've read Chapter 7 and have a firm understanding of its context.
- I have created all my social media accounts.
- I am familiar with the demographics of each platform.
- I have a strategy for the short-term promotion.
- I can list the communities I am a part of.

Chapter 7 Questions to Ask Yourself

1. Do I understand the objective of each social media platform?
2. Am I going to start promotion with __________ platform?
3. What does my short-term strategic plan consist of?
4. Am I going to measure success by…?

8

Copyright Law and the FCC

As with most of the chapters in this book, I'll open up this chapter by saying this is not a course on media law, and you will not be ready for the Bar examination by the chapter's end. What I am hopeful of is your understanding of basic media law, specifically copyright law and what you can and cannot do with elements of your podcast.

So, let's jump right in. Mark Twain said, "Only one thing is impossible for God: To find any sense in any copyright law on the planet." While you may not believe that now, I encourage you to revisit that quote at the end of this chapter. I didn't write the law, just teaching and passing it along.

Mike Flippo/Shutterstock.com

IMAGE 8-1 We the people

Three Myths

Before we get into some of the meat of this chapter, I want to debunk three copyright myths about podcasting that for whatever reason seems to be circling around as true.

Myth 1—The Thirty-Second Rule

The first myth is that it's OK to use copyrighted material (such as music, sounds, or other audio recordings) as long as you use 30 seconds or less. This rule does not exist. Using any portion of copyrighted material for a podcast without permission may be copyright infringement

Myth 2—The Credit Rule

The second myth is that it's okay to use copyrighted materials as long as you provide the appropriate acknowledgement to the copyright holder. However, this myth is also false. Giving credit to the copyright holder in your podcast does not mean you have permission to use the material. You still need to secure the rights.

Myth 3—The Nonprofit Rule

The last copyright infringement myth states if you are not making profit from your podcast you have carte blanche to use copyrighted materials in your podcast. Like the others, this myth is also false.

Copyright Law Q&A

In the United States, copyright law comes directly from the Constitution. Article I, Section 8, Clause 8 of the U.S. Constitution gives the basic understanding for how we developed our modern copyright law. It states, "*The Congress shall have Power … To promote the Progress of Science and useful Arts, by securing for limited Times to Authors and Inventors the exclusive Right to their respective Writings and Discoveries.*"

create jobs 51/Shutterstock.com

IMAGE 8-2 Copyright law

Copyright law has evolved over the years from its original meaning, and the current law was adopted in 1976. This version of copyright law out reaches almost all state laws, making copyright law virtually federal law.

This is not a bad thing, and it gives a certain set of rights to the copyrighted materials author, or the owner of the copyright. Under copyright law, these exclusive rights are recognized:

- Reproduce the work
- Prepare and create derivative works
- Publicly distribute the work
- Publicly perform the work
- Publicly display the work
- Publicly perform a digital sound recording

So What Exactly Can Be Copyrighted?

Good question, according to copyright.gov, "*Copyright, a form of intellectual property law, protects original works of authorship including literary, dramatic, musical, and artistic works, such as poetry, novels, movies, songs, computer software, and architecture.*"

Furthermore, according to copyright.gov, "Copyright does not protect facts, ideas, systems, or methods of operation, although it may protect the way these things are expressed."

There have been and there are going to be cases in the courts contesting previous rulings on what can be copyrighted and what cannot. However just from the material covered to this point in this chapter, can a podcast be copyrighted? Absolutely. How about that logo you created and put at the beginning of your podcast? 100% yes.

Song_about_summer/Shutterstock.com

IMAGE 8-3 Your podcast and logo can be copyrighted

Is There a Difference between Copyright and Patent or Trademark?

There is a difference. Copyright protects original works created by the original author. If you recall copyright does not protect facts, ideas, systems, or methods of operation. Patents are designed to protect inventions or discoveries. A trademark protects words, phrases, symbols, or designs.

When Is My Work Copyrighted?

Another good question. The answer is the moment it is created and fixed. The following is directly from copyright.gov and addresses when and how a work is fixed.

> *Copyright is a form of protection provided by the laws of the United States to the authors of "original works of authorship" that are fixed in a tangible form of expression. An original work of authorship is a work that is independently created by a human author and possesses at least some minimal degree of creativity. A work is "fixed" when it is captured (either by or under the authority of an author) in a sufficiently permanent medium such that the work can be perceived, reproduced, or communicated for more than a short time. Copyright protection in the United States exists automatically from the moment the original work of authorship is fixed.*

Just to break that down a little bit, you are the original author of your podcast, and the moment you hit the record button, your original work is fixed in a tangible form of expression. I'm going to assume you are human and have put a minimal degree of creativity into your podcast. Therefore you have successfully completed the first half of the above statement.

To be clear, to be fixed means to record your audio/video works to a storage device that can then be shared, uploaded, watched, edited, etc. Once your original work has made it to this point it is fixed. And therefore it is protected under copyright law. Again, so is everyone else's. To avoid copyright infringement the very best practice is to be the original author of all content on or related to your podcast.

Do I Need to Register with the Copyright Office to Be Protected?

Yes and no. I knew things were going to get a little gray somewhere in this chapter. I'll explain. No, you do not need to register with the copyright office, it is strictly voluntary. The moment your work is fixed it is protected by copyright law. Yes, you will have to register your original works if you ever wish to bring a potential lawsuit for infringement of work within the United States.

PX Media/Shutterstock.com

IMAGE 8-4 There are reasons for and for registering your original work with the copyright office

Additionally, you may want to register with the US copyright office in order to have it in public records. Also, if it ever came to it, registered works could be eligible for statutory damages and attorney's fees in the case of successful litigation.

I've Heard I Can Mail a Copy of My Original Work, and by Doing This It Is Registered as Copyrighted. Is This True?

I've heard the same rumor, not sure who started that. Anyhow, the idea is that by having your original works date stamped by the US postal office would establish the specific time that your works were created. This is often referred to as the "poor man's copyright" and while the idea seems solid enough, there is nothing in copyright law that supports this, and therefore this in no way registers your original work.

How Do I Register My Original Works with the U.S. Copyright Office?

Like many things in life (like researching for writing a book) things are easier in the age of the Internet. Anyhow, registering with the US copyright office is no exception. At the copyright registration site—www.copyright.gov/registration/—

you are going to want to navigate to the Motion Pictures portal. Copyright.gov defines motion picture as the following:

> *Motion Pictures are works that contain a series of related images that are intended to be shown with a projector, digital display, or other device. When the images are shown in successive order, they create an impression of movement that is perceptible to the eye. Examples of motion pictures include movies, television shows, video games, animations, and similar types of works.*

Your audio podcast or your audio/video podcast falls into this category. It is as simple as creating an account with a username and a password, and e-filing your work. You can still register by sending through the U.S. post office, however it is quicker and cheaper to e-file.

Fees associated with registering with the U.S. copyright office are going to vary, however, as of the writing of this book, the standard fee is $65.

Can I Find Out Who Owns a Copyright?

Yes, the U.S. copyright office will perform a search for ownership of copyright for all records from January 1978 to the present date. There is a fee associated with the search; however, there is no cost if an individual is willing to conduct the search themselves at the copyright office.

Wanan Wanan/Shutterstock.com

IMAGE 8-5 There will be a fee associated with the U.S. copyright office conducting a search for you

How Long Is the Copyright Good for?

This is a two part answer. The first part states that the copyright for a single original artist, the life of the copyright is for the duration of the lifetime of the original artist plus seventy years. If there is a group associated with the copyright, the life of the registration is the duration of the last living person of the group plus seventy years. This applies to all original works created after January 1, 1978.

How Much of Someone Else's Work Can I Use Without Getting Permission?

Under the fair use portion of the U.S. copyright statute, it is possible to use someone else's work whether it be quotes, or notes of a song. However there is no specific wording to the statute on how many words or how many notes can be used. I am of the opinion that you shouldn't wade into those waters, and avoid, completely, the use of someone else's work.

What If Someone Has Infringed on My Copyrighted Work?

As stated directly on copyright.gov:

> *A party may seek to protect his or her copyrights against unauthorized use by filing a civil lawsuit in federal district court. If you believe that your copyright has been infringed, consult an attorney. In cases of willful infringement for profit, the U.S. Attorney may initiate a criminal investigation.*

We are going to shift gears here. It is important that we briefly study FCC law: what the FCC is, what it regulates, and what you can and cannot do.

Federal Communication Commission

As stated directly on FCC.gov, the Federal Communications Commission (FCC) does the following:

The Federal Communications Commission regulates interstate and international communications by radio, television, wire, satellite and cable in all 50 states, the District of Columbia and U.S. territories. An independent U.S. government agency overseen by Congress, the commission is the United States' primary authority for communications law, regulation and technological innovation. In its work facing economic opportunities and challenges associated with rapidly evolving advances in global communications, the agency capitalizes on its competencies in:

- Promoting competition, innovation and investment in broadband services and facilities
- Supporting the nation's economy by ensuring an appropriate competitive framework for the unfolding of the communications revolution
- Encouraging the highest and best use of spectrum domestically and internationally
- Revising media regulations so that new technologies flourish alongside diversity and localism
- Providing leadership in strengthening the defense of the nation's communications infrastructure

Mark Van Scyoc/Shutterstock.com

IMAGE 8-6 The FCC has many roles in all 50 states

No matter what Eminem says, the FCC is not designed to prevent you from expressing your opinion, but rather to do so in the most equal competitive manner possible. I encourage you to read those bullet points again so as to get a complete understanding of the objective of the FCC.

Also know by design, the FCC is built to evolve with changing technologies. On June 19, 1934, the passing of the Communications Act officially founded the FCC, and they had zero idea we would one day create podcasts and stream through a variety of channels. Yet by the objectives of the act, it's able to regulate new technologies and keep the playing field level.

Chapter 8 Terms

Copyright Intellectual property that protects the original author once the works are fixed to a tangible medium.

Federal Communication Commission Regulates interstate and international communications by radio, television, wire, satellite and cable in all 50 states, the District of Columbia and U.S. territories.

Patent Exclusive right that is granted for an invention.

Three copyright myths Thirty-second rule, the credit rule, and the nonprofit rule.

Trademark A type of intellectual property made up of a recognizable sign, design, or expression that works to identify a product or service from a particular source and distinguishes them from all others.

Chapter 8 Checklist

- I've read Chapter 8 and have a firm understanding of its context.
- I understand the difference in copyright, patent, and trademark.
- I know the three myths of copyright law.

Chapter 8 Questions to Ask Yourself

1. When is original work copyrighted?
2. Why would I register my original work with the copyright office?
3. For how long is original work copyrighted?
4. What are the functions of the FCC?

9

Keeping It Going

To this point, I hope you are enjoying the journey and creating some excellent content. It has been my experience with both my own podcast, and the podcasts of my students that several things surface that need to be addressed so you can better deal with them if and when they arise for you. The first, and we briefly discussed this in Chapter 1, is pace. Controlling and maintaining a steady pace for your podcast is vital. You may feel as though you want to record a dozen episodes in the first week. Let me say that I am happy for your excitement and let me encourage you not to record a dozen episodes in the first week.

By now you've noticed several quotes throughout the book, so here's another one, by Steve Kaczmarski, "The No. 1 cause of burnout is doing the same thing over and over again and not seeing results."

fizkes/Shutterstock.com

IMAGE 9-1 Quick burnout is the number one reason for podcasts failing

I'll write this again in case you missed it, burnout is the number one reason for podcasts being unsuccessful. So, real quick, let's do a recap of the roles you are going to doing as you get you podcast off the ground:

- Scheduling guests
- Being talent on your podcast
- Script writing
- Audio and video editor
- Studio engineer
- Creating all social media posts
- Engaging with your social media community
- Study analytics

You've got a lot to do with each episode, pace yourself. Create a schedule you can live with and give yourself plenty of time to do each of the jobs you need to do. Resist the temptation to get in over your head at the beginning and bite off more than you can chew. If you don't the result is usually burnout, and with that the podcast usually dies out.

The second, and one of the more difficult things to overcome in your podcast journey is staleness. The Cambridge Dictionary defines staleness as, "the condition of having lost interest in what you are doing because you are bored or are working too hard." Go ahead and take a second to read that again. Let me assure you, you are going to run into feelings of being bored and feelings of working too hard. The latter being most likely the first. Losing interest in what you are doing is going to rear its ugly head as well.

Here are some tips to defeating your boredom:

- Being a student hopefully you are using your logic-thinking left side of your brain, and the creative-thinking right side of your brain. If you're bored with your creativity, perhaps it's time for some philosophical readings. They'll exercise your logical thinking and rest your creative thinking.
- Get outside, seriously. Walk away (not permanently) from your project and get outside, even if it's ice cold. It's good for the soul and for reestablishing the reason you got into podcasting to begin with.
- Similarly to getting outside, set time aside for yourself. Plan on off time. It's good for motivation.

KieferPix/Shutterstock.com

IMAGE 9-2 Sometimes all you need to do is find a little motivation to keep your project going

The above mentioned tips apply directly to working too hard. If you're a student, you have classes, homework, regular work, a social life, and probably half a dozen other things. I get it, you are busy and working hard. But you get through these obstacles because you hopefully are motivated by graduating and getting a better paying job. Establish your motivation for beginning your podcast in the beginning and let that remain your overarching purpose. If you focus on that, perhaps you won't be tempted to focus on boredom, or working too hard.

Staying Motivated

I know as well as anyone the struggle of maintaining motivation long after the new has worn off and the shine has faded. In fact writing the book you're reading at this moment I may have exercised a few of the methods for keeping motivated. You may have some preferred methods, but I'll share six of mine.

1. Visualize yourself being successful. You may be one of those who began your podcast with zero purpose of fame, but rather for the love of your topic. Great, however when that love runs low, you need to visualize yourself as being successful and maybe even famous. Believe it or not, these visualizations will allow your brain to release dopamine, the "pleasure hormone". Soon your brain's association with success and pleasure could allow you the mental tools to continue on.
2. Somewhere in the planning and season-mapping process, you more than likely developed goals, and that's great. You should have goals. I encourage you to examine those goals. Do they look something like having six digit followers, YouTube royalty deposits every month, and deciding if you want to quit your day job and go full time podcasting? If so, that's also great. That also takes time, and lots of it. If you keep your eye on the end goals you'll most likely lose motivation and your podcast will die on the vine. I encourage you to set smaller, more manageable goals, and lots of them. How about the first goal being to record your first episode, edit your first episode, and publish your first episode. That's a manageable goal, and once reached, you'll find encouragement to set and reach other manageable goals. At first you might believe you don't need this. Believe me, at some point, you will.

patpitchaya/Shutterstock.com

IMAGE 9-3 Be SMART when setting your goals

3. Building on the idea of setting goals, it really does matter how you perceive yourself. By that I mean do you set goals with a negative or punishing result if the goal is not met, or do you set your goals with a positive, rewarding outcome if met? It really does make a difference. There will be scholars who would disagree, and I know operant conditioning theory states otherwise; however, we are a one person team that needs to find ongoing motivation quite frequently, and positive reinforcements can do wonders. Oh, and by the way, operant conditioning is just a theory.
4. Stay consistent in your preparation. Sure, some details are going to change in the guest scheduling, and in the script writing, and even in the editing. You are going to discover what works best for you. However, avoid that little voice in your head that's going to encourage you to cut corners. Don't do it. Stay consistent. Be detailed, be complete in your work, and do it consistently.
5. Get a friend on board. Sometimes it can get really difficult motivating yourself. Keeping yourself accountable is hard, and some of us are not very good at it. If you have a friend or two on your podcast journey it can make the trip a lot less bumpy. Just make sure this friend has similar interests and goals and the two of you can be honest about holding each other accountable.

Dean Drobot/Shutterstock.com

IMAGE 9-4 Working together with a friend can go a long way to keeping you motivated

6. Have a list and keep it close by. Physically write a list of your goals and your purpose for doing your podcast. I encourage you to write this on paper, not on your phone, and keep this list with you. Make this list when motivation is high, and you feel on top of the world. Then, when motivation is low, and the thought of quitting enters your mind, read your list. You'll remember those feelings of excitement when you wrote your list, and it's just possible a little of that spark will return, and motivation will begin to grow.

Motivation, persistence, things we've discussed along this journey are some tips and tricks you can use along the way to keep yourself and your podcast on track. However, there is one thing that you must apply every day in order to make your podcast a success, and that is hard work. It wouldn't be right to not use a quote here, so I'm going to quote Vince Lombardi, "The man on the top of the mountain didn't fall there." Hard work doesn't always mean success, but success never happens without hard work.

10

Troubleshooting Your Problems

Nothing is more frustrating than having a great interview only to find out you recorded nothing. Or a constant buzz in your headphones that cannot be ignored and will not go away. Or even worse, lighting that makes the talent look yellow.

There is little to no doubt you are going to run into problems when you record your podcast, just prepare for it now. Because I've been running into technical issues since the beginning of working in a studio, I wanted to put a chapter together that introduces the problem, and give a workable solution. When you run into problems, I hope you find your solution here.

docstockmedia/Shutterstock.com

IMAGE 10-1 Troubleshooting

Audio Issues

Buzz can be heard in earphones—If you remember back in Chapter 3 we discussed balanced and unbalanced cables. If you are using an unbalanced microphone cable, and the length of the cable is longer than 10 feet, you may be picking up a buzz due to distance. More likely you are using a balanced cable so length isn't an issue. Check to make sure your microphone cable isn't overlapping other cables, electrical cables, or even power strips. Your balanced cables will pick up distortion if in contact with other signal carrying cables.

No sound—A common problem, usually there is an easy to overlook solution. Check to make sure all cables are properly plugged in to the mics and to the audio board. If that is correct, check to make sure your mics are not muted at the board. If they are not, check volumes on the board.

Too much bass—Another common problem, and this has an easy solution. Most of the time it is a mic too close to a sound source. For example, if the host is too close when speaking, there will be a bass effect. The easiest solution is to move the mic until the bass sound has been removed.

Too much natural sound—This could result from one of two problems. The first one may be as simple as the mic is too far from the talent. Move the mic closer. The second (and the more probable causes) you may have a lack of sound absorbing panels or tiles. As we discussed in Chapter 3, in order to get the acoustics right in your studio you'll need sound absorbing panels. This will lower natural sounds and vocals bouncing off the walls.

Plosives—From time to time, this is a problem in just about every studio. The "p" and "b" words can often distort the mic if only for a split second. The audio recording picks this up, and messes up your otherwise clean audio. The easiest solution is to have pop filters for every microphone. These screen filters will distort the "p and "b" words and that distortion will never hit the microphone, therefore eliminating the problem.

paffy/Shutterstock.com

IMAGE 10-2 Pop filters will eliminate your plosives

Foot noise—In an ideal world there would never be a concern for foot noise because no one moves around during the recording, right? Not exactly, sooner or later it's going to happen. Adding a shock mount will reduce the mics picking up these sounds. A shock mount works by creating acoustics between the mics and the mic stand.

Distortion—This is heard during playback and it is an electrical static being picked up in your recording. This is probably a gain issue and it's most likely turned too high. Reduce gain on that channel, and that can be found in the audio tab of your digital interface that your device is hooked to, or on the board's screen.

Picking up hiss—This can be misdiagnosed as distortion, and it could sound somewhat similar. More times than not, it is the microphone. Saving money and getting the cheaper microphone can really cost more in the end, avoid this in the beginning and get quality mics at the start.

spaxiax/Shutterstock.com

IMAGE 10-3 Don't pay more in the end; begin with quality mics while designing your studio

Only getting sound out of one speaker—It is a rare occurrence, but it can happen. When you are only getting sound out of one speaker, more than likely the left speaker, you are recording in mono and not stereo.

Video Recording Issues

Lighting color is different on different cameras—Cameras need to be white balanced. Place a plain white piece of paper or foam board in the camera view. Zoom in on the white board until the screen is fully white. Then focus the camera and wait about three to five seconds. Zoom out and the camera should be white balanced. Repeat this with each camera until all cameras are the same color temperature.

Shadows on Talent—A common problem that is fairly easy to address. The problem is your lighting and their location. Adjust tilt, but do this in small increments. If the tilt doesn't solve your shadow problem, try raising your lights. Make sure your lighting source is equal on all sides.

Video cannot be opened by my editing software—Another common issue. Some recording formats are not compatible with all editing software. The easiest solution is to download and install VLC. You can open your recording inside the VLC software dashboard and convert the file to one that is compatible with your editing software. Mp4 is a common file type that is editable in most editing software programs.

monticello/Shutterstock.com

IMAGE 10-4 It's a good idea to download VLC before you run into video file type issues

Audio/video are out of Sync—A frustrating problem to have to deal with. There are two solutions here, first is to sync audio/video files in your editing software. This is done in the timeline panel. Simply select the files, unlink the audio and video files and adjust your audio until it matches with the video. The second solution is to use VLC (I'm kind of a fan of this free software). Once your video is opened in VLC there is an audio sync manual option. Select this and your audio and video should sync.

Talent goes out of focus during recording—When your camera is set to automatic focus, it will temporarily blur when talent or guests move. The camera is trying to focus during this movement, causing blur. The easiest solution, if this is a common problem, is to shift the camera to manual focus.

mapman/Shutterstock.com

IMAGE 10-5 Your camera if set to auto focus can blur as guests and talent move

Movements within the video are choppy—This is a frame rate issue. The cause is most likely a too low of a frame rate. Frame rate is the frequency at which images are captured. The higher the frame rate, the smoother the faster the action is going to look.

Video quality is bad—This could be a number of issues. First, look into the device you are recording your video to. Is it a computer? SD card? Make sure your computer has the specifications required to capture good video. It could be something as easy as deleting your computer's cache. If you are recording to an SD card, make sure it is a class 10 (indicated by a circled 10 on the card). This means the card is capable of being written to as fast as the camera is recording.

Sean Liew/Shutterstock.com

IMAGE 10-5 You made it!

Glossary

Adaptive high bit rate Allows for your video to adjust to whatever Internet connection your viewer has and your video will automatically adjust for seamless viewing.

Analogous color scheme Uses three hues that are located next to one another on the color wheel. Usually, this scheme involves one dominant color, either a primary or secondary color, and then two supporting colors.

Analytics Information resulting from the systematic analysis of data or statistics.

Anchor.fm A one-stop shop for all your audio only, or audio and video podcasts.

Audio board A board capable of taking multiple input signals such as microphones, and merging them together to create one signal that can be sent to your recording device for later use.

Audio compression The technique of reducing the dynamic contrast within an audio waveform whereas audio format compression is the computational technique of reducing the file size of an audio file with minimizing a reduction in quality.

Audio editing The process of manipulating recorded sound, whether it be an analog or digital medium, for the purpose of greater functionality, increased efficiency, and aural aesthetics.

Audio/video checklist A list of actions that should be taken before each recording to assure you are getting the highest recording quality possible.

Balanced audio An audio cable that contains three wires, one positive audio signal, one negative audio signal, and one ground.

Cohost The second of the two individuals leading the podcast.

Community Members of a social media platform that share common interests.

Complementary colors Using two colors that complement each other in harmony. Usually opposites attract, and functioning complementary colors are red and green, orange and blue, yellow and purple.

Condenser microphone When you speak into a condenser mic, your sound waves hit a diaphragm. In the condenser mic, the diaphragm is actually one of two metal plates. As the diaphragm vibrates, the distance between these two metal plates shortens and expands. It is during this movement that charge is created and the condenser mic converts this charge into electric current, and moves it out of the mic to an amplifier or to a recorder.

Conversational-style podcast Format where a host and a cohost converse over a variety of topics.

Copyright Intellectual property that protects the original author once the works are fixed to a tangible medium.

Director This is the person who is in charge of the final product. They make decisions on just about every aspect of the show.

Dynamic microphone Designed with a diaphragm that absorbs and moves with incoming sound waves. An attached coil also moves around a magnet creating a magnetic field. As the coil moves through this magnetic field, electric current flows through the coil. The electric current moves through the microphone, to an amplifier or to a recorder.

Educational-style podcast The host and/or cohost focus on one particular element of their educational topic and each episode focuses on a different, related topic.

Engagement Term for actions that measure to what extent your audience interacts with your posts.

Episode mapping Creating a list of shows for the entire season.

Federal Communication Commission Regulates interstate and international communications by radio, television, wire, satellite, and cable in all 50 states, the District of Columbia and U.S. territories.

Floor director The person who monitors camera shots, lighting on talent to avoid shadows, and works directly with the director.

Guest A knowledgeable individual brought onto the show to speak about a specific topic.

H.264 A video compression standard known as MPEG-4 Part 10, Advanced Video Coding, MPEG-4 AVC, or AVC video.

Hand tool Allows you to grab and move clips, graphics, and other elements to your timeline.

High-pass filter An EQ curve that is used to reduce or remove frequencies below a specific cutoff frequency in an audio signal.

Host The individual leading the podcast.

Interviewing Asking a series of questions to a guest in order to get opinions or experiences.

Interview-style podcast Format where host and cohost have a variety of guests to discuss certain topics. Topics can vary, however the overarching theme is consistent.

Intro The beginning of your podcast that gives your viewer or listener an introduction to what they can expect from that episode of your podcast.

Modern font It is as the name implies, and it meant to give a modern, distinctive, forward thinking image.

Monochromatic The idea of using one color with different hues, or shades within that color's spectrum.

Nonfiction podcast This is a podcast that uses nonfiction topics and discusses and informs listeners on a wide range of topics. Unlike the educational format, this format allows the host to switch topics for each episode.

Nonverbal cues Is the transmission of messages or signals through the absence of speaking such as eye contact, facial expressions, hand gestures, posture, or the use of objects and body language.

Novelty/display fonts These are meant to be fun, wacky, and eccentric.

On-air presence The fact or the condition of being aware of the image you are creating of yourself while in front of the camera.

Panel-style podcast The host and co-host have a different panel of experts to discuss specific topics. Guests are knowledgeable in the field of the discussion.

Patent Exclusive right that is granted for an invention.

Pen tool Allows you to draw straight, or curved paths.

Persistence Doing something despite difficulty or delay in achieving success.

Plosives When the air flow from the mouth is momentarily blocked by the lips closing, and then followed by a rush of air.

Podcast A digital file, usually an audio file, that has been uploaded to the Internet so that a listener can download the file onto a digital device for listening.

Prep time All the combined time spent preparing for the recording of a podcast.

Program panel The portion of the editing software that shows you a preview of your project.

Project panel The portion of the editing software interface that contains bins, searches for files on your computer or server, has links to effects.

Promotional posts Social media posts that you set a budget for, and they are designed to reach a far greater audience than an organic post.

Razor tool Allows you to cut or split clips to isolate the parts of video you wish to keep or delete.

Rectangle tool Allows you to draw a variety of shapes.

Ribbon microphone Works by using a thin vibrating ribbon of metals, thinner than the plates in the condenser mics, that move between the poles of a magnet. These mics do not use diaphragms and coils; instead, the movement between the poles generates the electric current, which moves through the microphone to an amplifier or to a recorder.

Ripple edit tool Allows you to lengthen or shorten a clip while moving the rest of your timeline to accommodate.

Sans serif fonts The term "sans" means without, and in this case it is letters without flags or without tags.

Script fonts Cursive fonts, these are somewhat decorative fonts, but they do have purpose and have been in use for many years.

Scripting for episodes Creating areas of focus as pointers to use throughout the recording of your episode.

Scripting for intro Brief introduction to each episode with the purpose of letting your viewers know immediately what show they are tuning into, let viewers know what they can expect from your podcast, give an extended tagline, identify the show's host or hosts, let the viewers know what to expect on this episode, and to provide a wrap-up.

Selection tool Allows you to do basic selections, edits and movement.

Serif fonts Basic type that has a little character to certain letters. This character is referred to as tags, or flags.

Sibilances High-frequency hisses that occur when using syllables such as S and sometimes T and Z.

Slip tool Allows you to change which part of a clip to display while keeping it the same size and length on the timeline.

Social listening The act of observing what the discussion is within your community.

Solo Show Podcast format where you are the host and there isn't a guest.

Source panel The portion of the editing software that allows you to view clips individually, and stores multiple clips in one place.

Talent The person or persons in front of the camera hosting the podcast.

Technical director The person who runs the audio board, the switcher, and operates graphics during live recording.

Tetradic (double complementary) scheme The most complex of all of the color schemes because it uses four colors which are organized into complementary pairs.

Text tool Allows you to type text over your video.

Theater podcast This is where the host or host and cohost have an opportunity to share their short fiction stories.

Three copyright myths Thirty-second rule, the credit rule, and the nonprofit rule.

Three-to-one posting ratio The practice of posting three posts related to your podcast, or the topic of conversation within your community to every one post promoting your podcast.

Timeline panel The portion of the editing software interface that stacks audio and video clips for arranging, and editing.

Track select forward tool Allows you to select a point on the timeline and move all clips that are before or after that point in either direction.

Trademark A type of intellectual property made up of a recognizable sign, design, or expression that works to identify a product or service from a particular source and distinguishes them from all others.

Triad This is a method of using three colors that are equal in distance from one another on the color wheel, forming a triangle.

TRS cable A balanced cable consisting of a tip, a ring, and a sleeve.

TS cable An unbalanced cable consisting of a tip and a sleeve.

Two F's Failure and frustration (prepare for them; they're just around the corner).

Unbalanced audio An audio cable that contains two wires, one is the signal and the other is the ground.

Video switcher Your studio cameras are hooked up into the switcher, allowing for the technical director to switch between camera angles during live recording.

Vodcast A digital file similar to a podcast; however, there is a video component accompanying the audio file to be watched as well as listened to.

XLR three-pin cable A standard audio cable, male-to-female connector. Pin 1 is ground, pin 2 is positive polarity mic audio signal, and pin 3 is negative polarity mic audio signal.

www.ingramcontent.com/pod-product-compliance
Lightning Source LLC
LaVergne TN
LVHW021930101126
852251LV00023B/261